YLP®

Young Learner's

MENTAL MATHS

Grade 1

Rajesh Singh

AF540056

CONTENTS

Published by:

YOUNG LEARNER PUBLICATIONS®

G-1A Rattan Jyoti, 18 Rajendra Place

New Delhi-110008 (INDIA)

Tel.: 011-25750801, 25820556, 25755559

Email: goodwillpub@gmail.com

gph.ylp@goodwillpublishinghouse.com

Website: goodwillpublishinghouse.com

Worksheet-1

Numerals and Number Names

1. Count and write the numerals and number names.

a)		1	One
b)			
c)			
d)			
e)			

2. Count and match the numerals and number names.

a)		5	Two
b)		7	Four
c)		2	Eight
d)		4	Seven
e)		8	Five

Worksheet-2

Forward and Backward Counting

1. Count forward and write the numbers in the boxes.

a)

1	2							9

b)

17		19					24

c)

29			32					37

d)

60				64			67

e)

92						98		100

2. Count backward and write the numbers in the boxes.

a)

7						1

b)

27			24				20

c)

45							38

d)

87					82

e)

99				95			92

Worksheet-3

Number Names and Numerals

Number name	Numeral
Seven	7
Eight	
Thirteen	
Twenty-five	
Thirty-four	
Forty-nine	
Fifty-one	
Fifty-eight	
Sixty-eight	
Sixty-six	
Seventy	
Seventy-eight	
Eighty-four	
Ninety-one	
Hundred	

Numeral	Number name
4	Four
9	
17	
21	
37	
43	
55	
59	
61	
64	
73	
76	
85	
94	
99	

Worksheet-4

Ones and Tens
After, Before and In Between

1. **Fill in the blanks.**

a) 23 ____ tens ____ ones   b) 49 ____ tens ____ ones

c) Thirty-six ____ tens ____ ones   d) Fifty ____ tens ____ ones

2. **Write the numeral for:**

a) 3 tens 2 ones ________   b) 5 tens 4 ones ________

c) 7 tens 6 ones ________   d) 9 tens 1 one ________

3. **Write the number name for:**

a) 2 tens 7 ones ________   b) 5 tens 8 ones ________

c) 8 tens 4 ones ________   d) 9 tens 6 ones ________

4. **What comes before?**

a) _____, 5   b) _____, 26

c) _____, _____, 76   d) _____, _____, _____, 100

5. **What comes after?**

a) 7, _____   b) 32, _____

c) 64, _____, _____   d) 97, _____, _____, _____

6. **What comes in between?**

a) 7, _____, 9   b) 34, _____, 36

c) 75, _____, _____, 78   d) 96, _____, _____, _____, 100

7. **What comes before and after?**

a) _____, 4, _____

b) _____, _____, 35, _____, _____

c) _____, _____, _____, 97, _____, _____, _____

Worksheet-5

Place Value, Expanded Form, Increasing and Decreasing Order

1. Write digit in ones (O's) and tens (T's) place.

a) 24: Digit in T's place ______, digit in O's place ______

b) 67: Digit in T's place ______, digit in O's place ______

2. Write the place values.

a) 43: Place value of 4 is ______ and place value of 3 is ______.

b) 81: Place value of 8 is ______ and place value of 1 is ______.

3. Write the expanded form.

a) 73 = ______ + ______   b) 85 = ______ + ______

4. Fill in the blanks.

a) The numeral with 5 at tens place and 8 at ones place is ______.

b) The numeral in which place value of 7 is 70 and 5 is 5 is ______.

c) The numeral with expanded form 40 + 8 is ______.

5. Put the correct sign (< or = or >) in the blanks.

a) 18 ______ 38   b) 57 ______ 25   c) 42 ______ 42

d) 59 ______ 57   e) 67 ______ 75   f) 91 ______ 91

6. Write in descending order.

a) 14, 67, 49: ______, ______, ______

b) 53, 56, 92, 48: ______, ______, ______, ______

7. Write in ascending order.

a) 92, 35, 21, 67: ______, ______, ______, ______

b) 28, 35, 31, 100, 53: ______, ______, ______, ______, ______

Worksheet-6

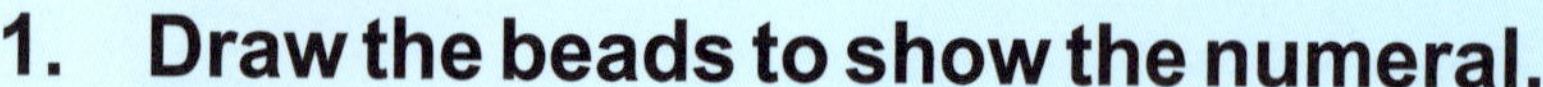

Abacus Time

1. **Draw the beads to show the numeral.**

a) 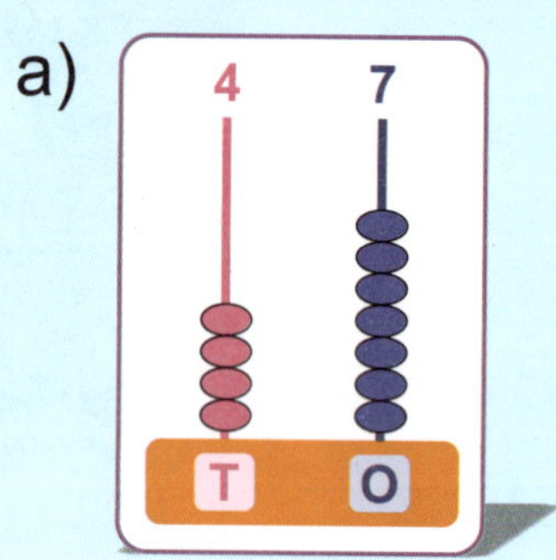

b) 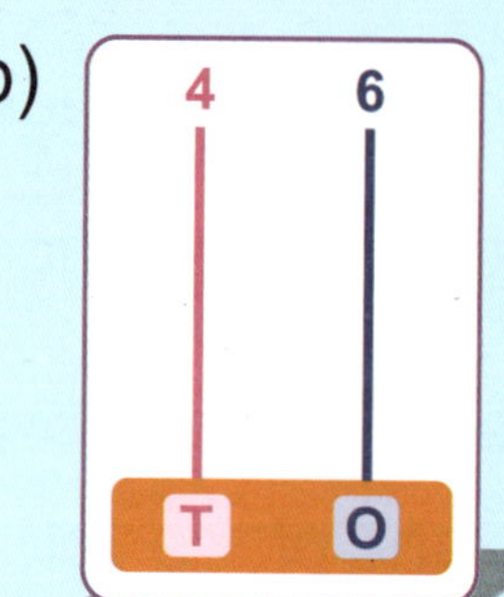

c) 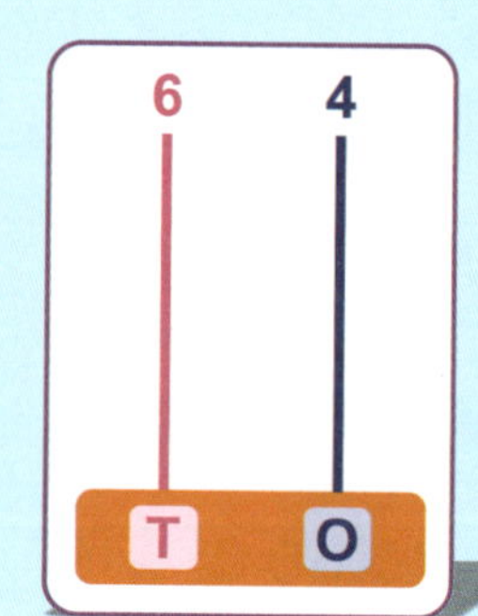

d) 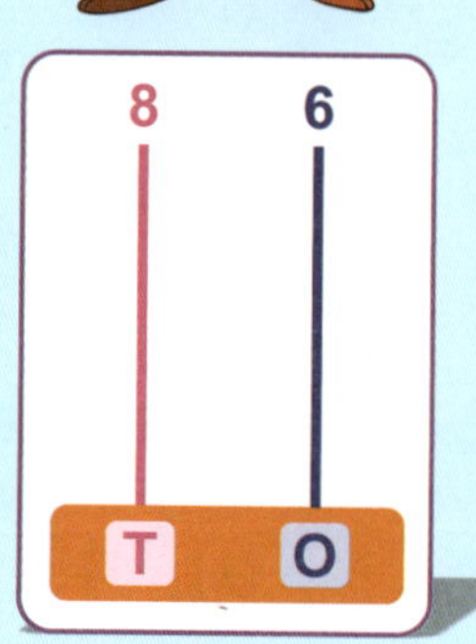

e) 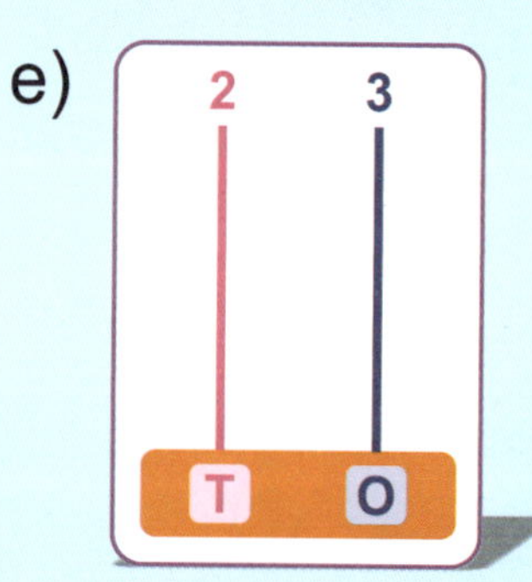

f) 

g) 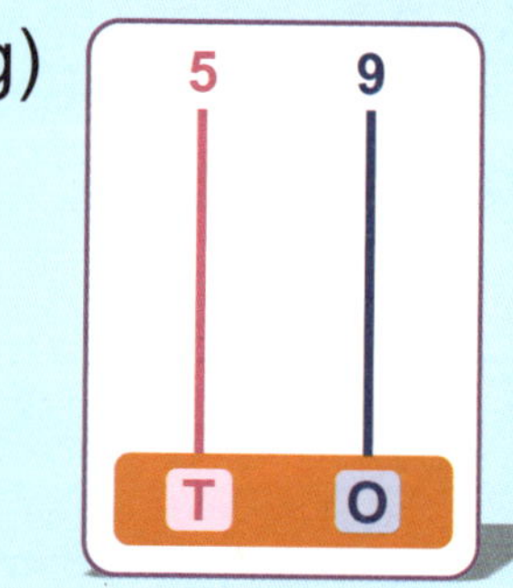

h) 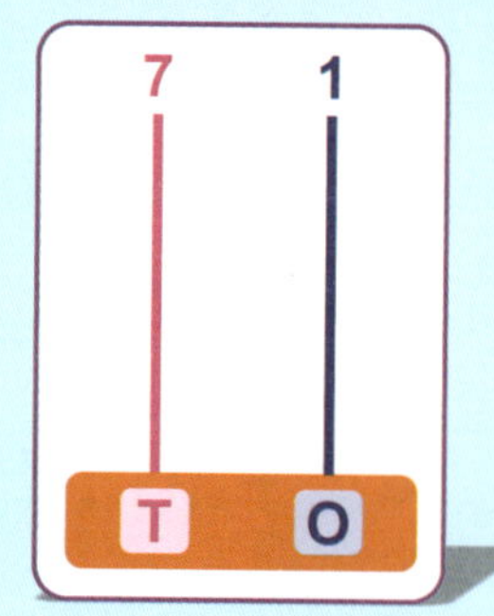

2. **Count the beads and write the numeral.**

a) 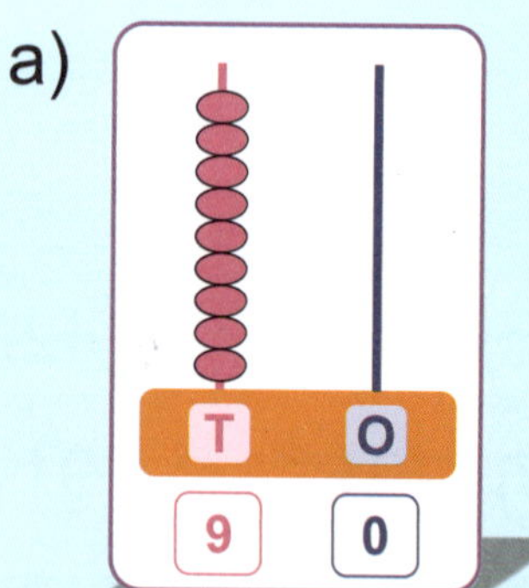

b) 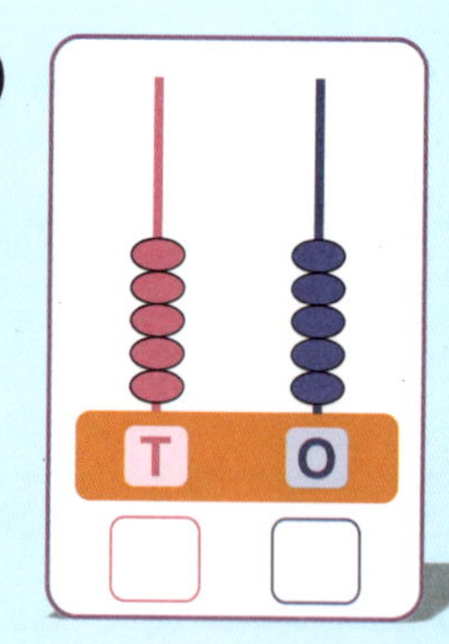

c) 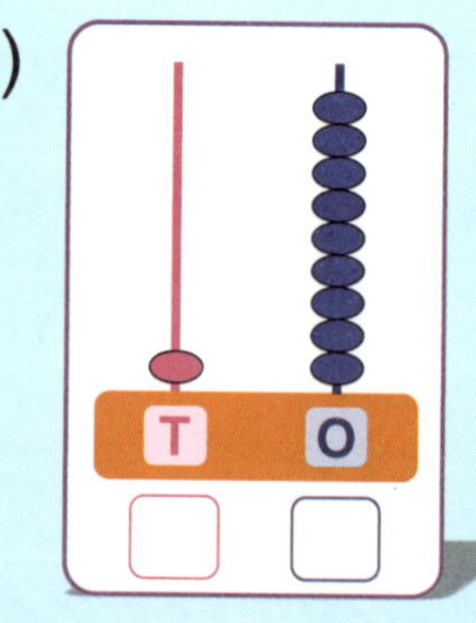

d) 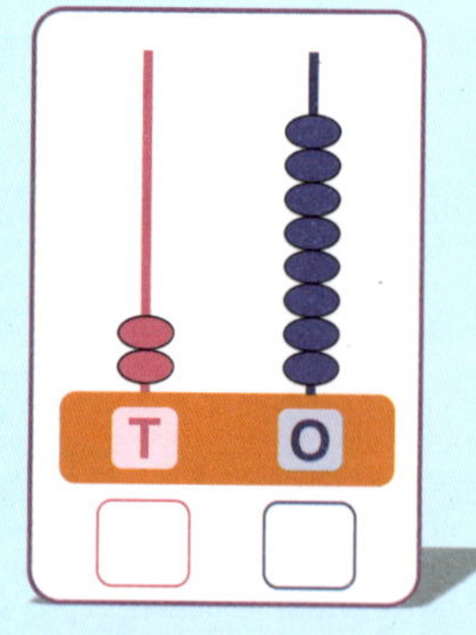

e) 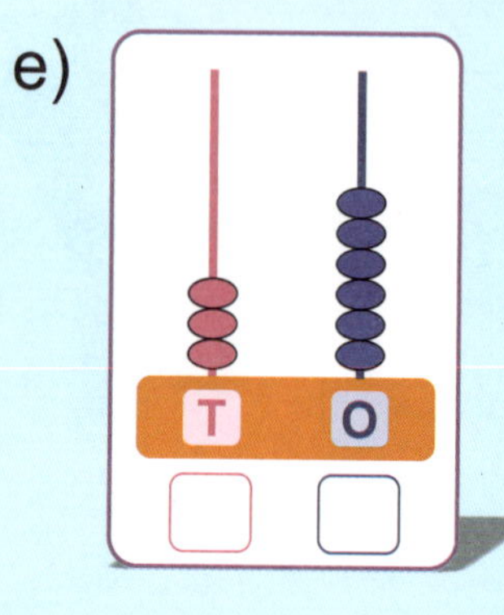

f) 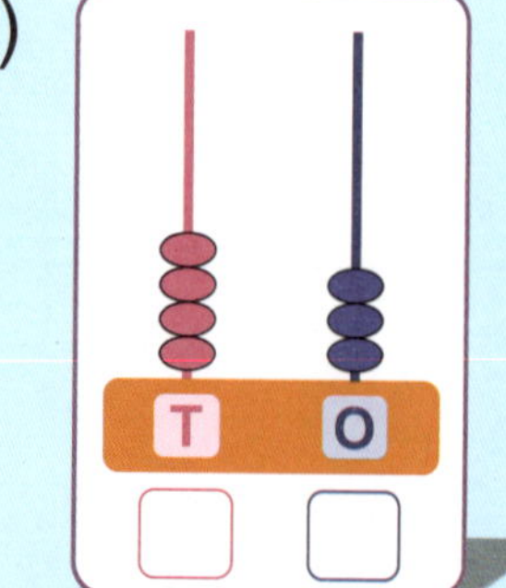

g) 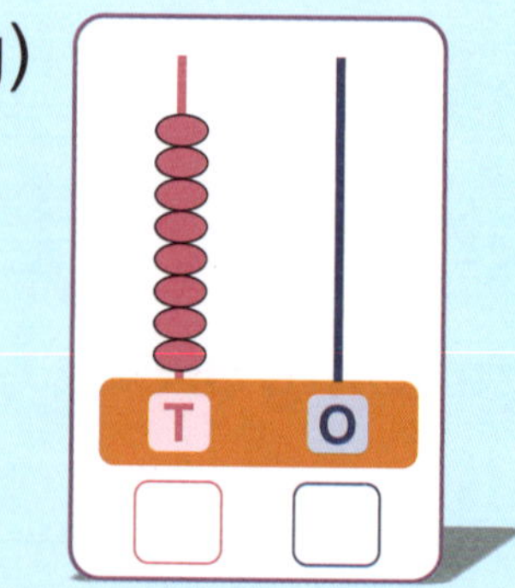

h) 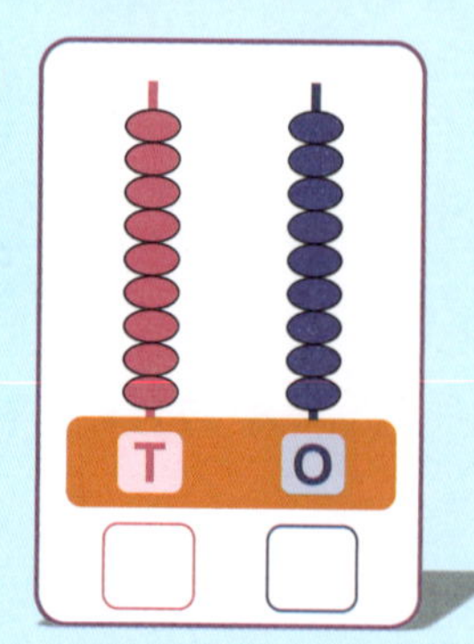

Worksheet-7

Smallest and Greatest Numbers Using Given Digits

1. **Make the smallest and greatest two-digit numbers using 3 and 6. Repetition is not allowed.**

   Smallest: ________________ Greatest: ________________

2. **Make the smallest and greatest two-digit numbers using 2 and 8. Repetition is allowed.**

   Smallest: ________________ Greatest: ________________

3. **Make the smallest and greatest two-digit numbers using 5 and 0. Repetition is not allowed.**

   Smallest: ________________ Greatest: ________________

4. **Make the smallest and greatest two-digit numbers using 0 and 9. Repetition is allowed.**

   Smallest: ________________ Greatest: ________________

5. **Make the smallest and greatest two-digit numbers using 4, 6 and 9. Repetition is not allowed.**

   Smallest: ________________ Greatest: ________________

6. **Make the smallest and greatest two-digit numbers using 7, 9 and 0. Repetition is allowed.**

   Smallest: ________________ Greatest: ________________

7. **Write all the possible two-digit numbers using the digits 4 and 9.**

   ____

   Smallest: ________________ Greatest: ________________

8. **Write all the possible two-digit numbers using the digits 1, 3 and 5.**

   ____

   Smallest: ________________ Greatest: ________________

Worksheet-8

Ordinal Numbers 1 to 10

1. Fill in the boxes.

Numeral	Number name	Ordinal number	Also written as
1	One	First	1st
		Second	
		Third	
4			
			5th
		Sixth	
	Seven		
8			
			9th
		Tenth	

2. Fill in the blanks.

Left	A	S	D	F	G	H	J	K	L	P	*Right*

a) Fourth letter from right is ________.

b) Third letter from left is ________.

c) K is ________ from right.

d) S is ________ from left.

e) L is second from ________ (right/left).

f) H is sixth from ________ (right/left).

g) G is ________ from right and ________ from left.

h) F is fourth from _____ (right/left) and seventh from ______ (right/left).

Worksheet-9

Addition by Counting

1

1	and	4	is equal to	5
1	+	4	=	5

2

  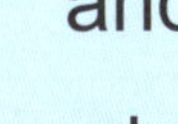  

	and		is equal to	
	+		=	

3

	and		is equal to	
	+		=	

4

   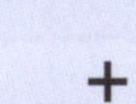  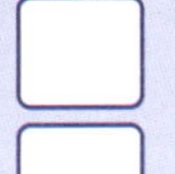 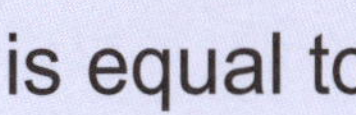  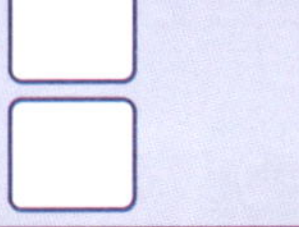

	and		is equal to	
	+		=	

5

  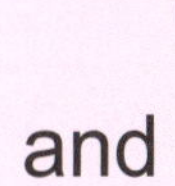 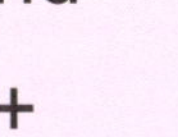  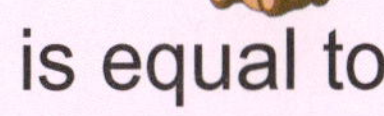 

	and		is equal to	
	+		=	

Worksheet-10

Addition by Forward Counting and Number Line

1. Write the numbers in the apples and add by forward counting.

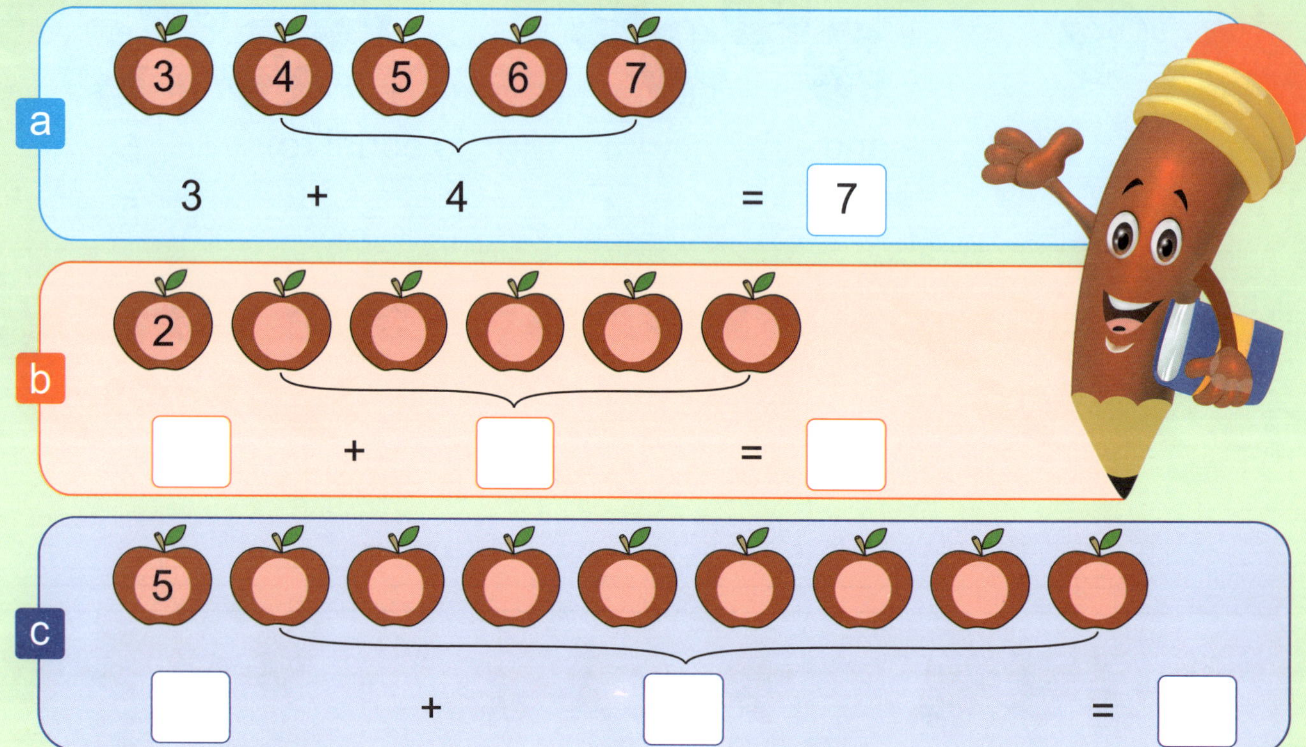

2. Add by using number line.

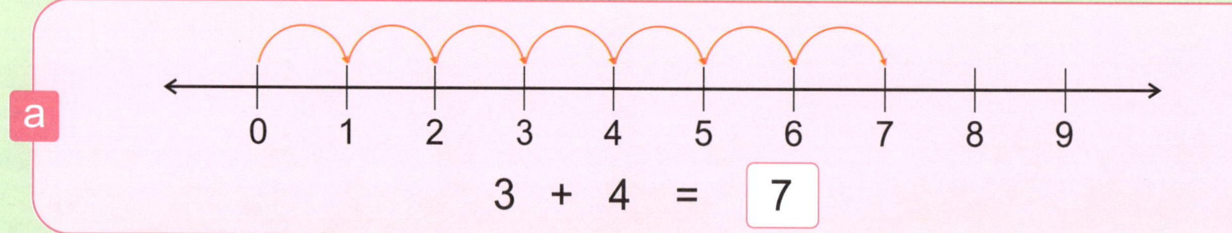

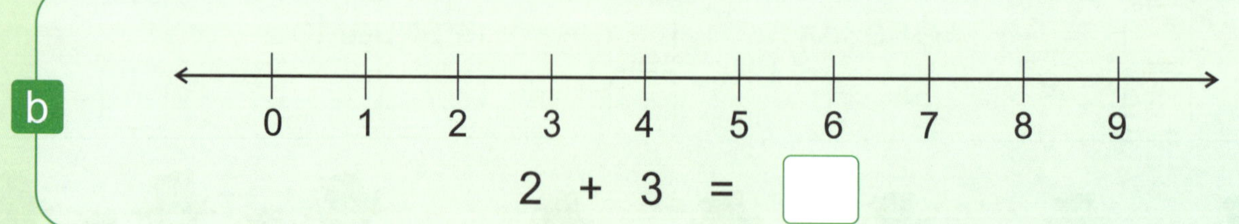

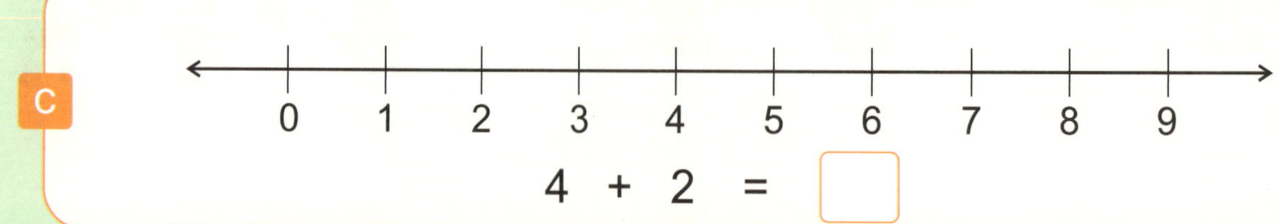

Worksheet-11

Addition by Drawing Lines and Regrouping

1. Add by drawing lines.

a) 3 + 5 = 8
| | | (1 2 3) | | | | | (4 5 6 7 8)

b) 4 + 6 = ☐

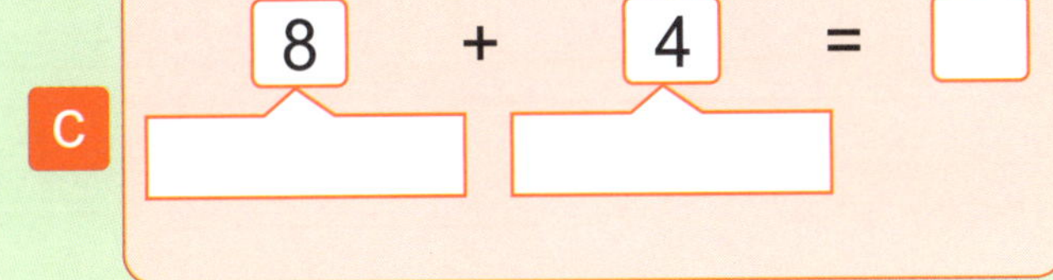

c) 8 + 4 = ☐

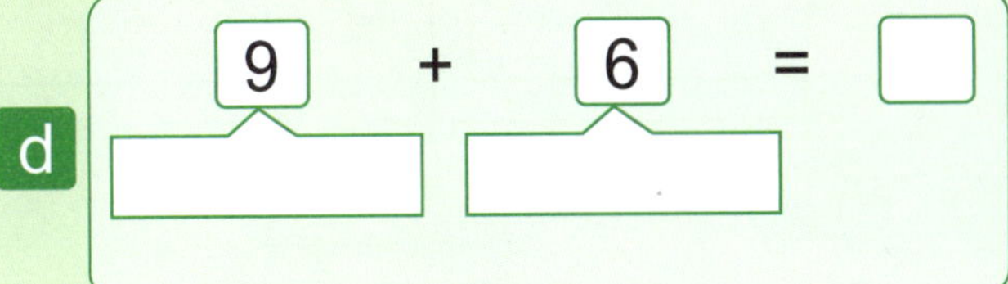

d) 9 + 6 = ☐

e)
```
    3   ← | | |  (1 2 3)
  + 2   ← | |    (4 5)
  ----
    5
```

f)
```
    4   ← ☐
  + 3   ← ☐
  ----
    ☐
```

g)
```
    5   ← ☐
  + 5   ← ☐
  ----
    ☐
```

h)
```
    5   ← ☐
  + 9   ← ☐
  ----
    ☐
```

2. Add by regrouping.

a) 37 + 21
= 3 tens + 7 ones + 2 tens + 1 one
= 3 + 2 tens + 7 + 1 ones
= 5 tens + 8 ones = 58

b) 41 + 28
= ____ tens + ____ one + ____ tens + ____ ones
= ____ + ____ tens + ____ + ____ ones
= ________ tens + ________ ones = ________

c) 35 + 54
= ____ tens + ____ ones + ____ tens + ____ ones
= ____ + ____ tens + ____ + ____ ones
= ________ tens + ________ ones = ________

Worksheet-12

Vertical and Horizontal Addition

1. Addition without carrying.

a)

4	3
+ 2	5

b)

3	3
+ 5	6

c)

2	2
+ 1	5

d)

7	6
+ 2	1

e) 32 + 52 = ________ f) 14 + 73 = ________

g) 17 + 41 = ________ h) 26 + 23 = ________

i) 16 + 14 = ________ j) 41 + 27 = ________

2. Addition with carrying.

a)

☐	
4	8
+ 1	7

b)

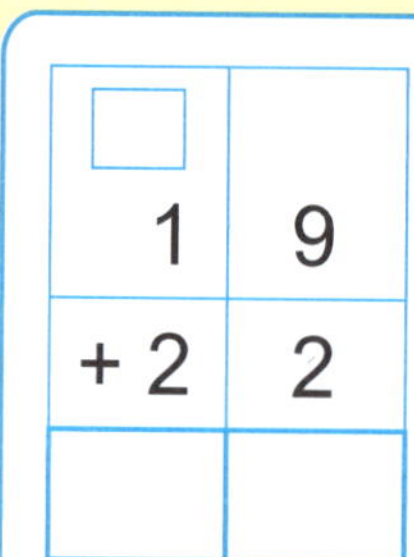

☐	
1	9
+ 2	2

c)

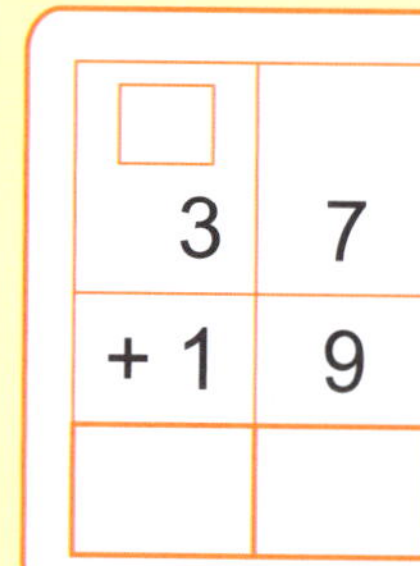

☐	
3	7
+ 1	9

d)

☐	
2	T
+ 4	7

e) 45 + 26 = ________ f) 29 + 59 = ________

g) 39 + 17 = ________ h) 27 + 38 = ________

3. Adding three numbers.

a)

☐	
4	5
1	3
+ 1	4

b)

4	4
2	0
+ 3	3

c)

☐	
1	4
2	3
+ 5	5

d)

☐	
1	8
4	2
+ 2	5

e) 48 + 34 + 13 = ________ f) 38 + 25 + 27 = ________

Worksheet-13

Addition Stories and Addition Facts

1. A store sold 6 shirts on the first day. 9 shirts were sold on the second day. How many shirts did the store sell in two days?

____

____

____

2. 39 kids were sitting in a hall. 32 more join them. How many kids are in the hall now?

____

____

____

3. There were 25 students in a music class. 12 more joined them. How many students are in the music class now?

____

____

____

4. There was a party at Helena's house. Her mother invited 37 guests. Her father invited 42 guests. Helena also invited 15 guests. How many guests were invited for the party?

____

____

____

5. Fill in the blanks.

a) 4 + 9 = 9 + ___ b) 5 + 3 = ___ + 5 c) 49 + ___ = 54 + 49

d) ___ + 39 = 39 + 23 e) 43 + ___ = 25 + ___ f) ___ + 44 = ___ + 21

g) 62 + 0 = ___ h) 99 + ___ = 99 i) ___ + 0 = 81

Worksheet-14

Subtraction by Counting

Subtract the following by counting.

1

3 ducks	1 moves away =	2 left
3 minus	1 =	2
3 –	1 =	2

2

☐ apples	☐ fall down	☐ left
☐ –	☐ =	☐

3

☐ birds	☐ fly away	☐ left
☐ –	☐ =	☐

4

☐ fish	☐ swim away	☐ left
☐ –	☐ =	☐

5

☐ planes	☐ fly off	☐ left
☐ –	☐ =	☐

Worksheet-15

Subtraction by Crossing Out

Subtract the following by crossing out.

5 – 2 = ☐

6 – 3 = ☐

7 – 5 = ☐

8 – 3 = ☐

10 – 5 = ☐

Worksheet-16

Subtraction by Backward Counting and Number Line

1. Subtract the following by backward counting.

2. Subtract by using number line.

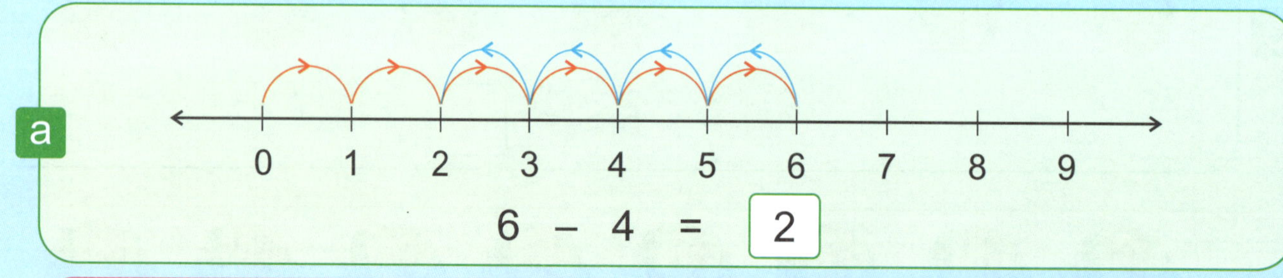

6 – 4 = 2

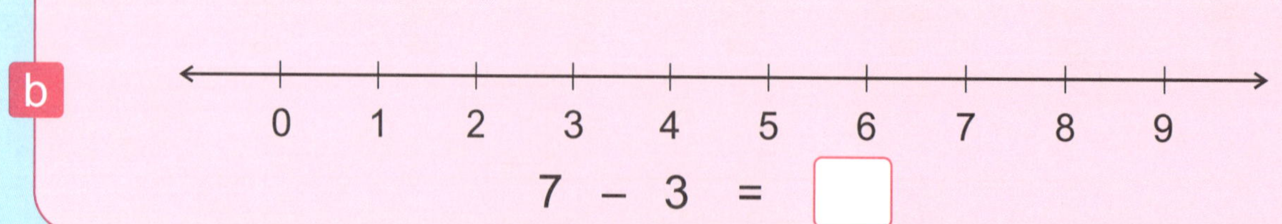

7 – 3 =

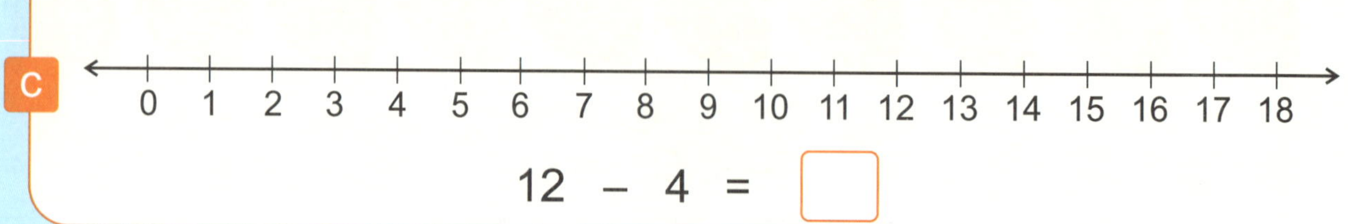

12 – 4 =

Worksheet-17

Subtraction by Drawing Lines and Regrouping

1. Subtract by drawing lines.

a) 7 – 4 = 3
/ / / / | | |

b) 9 – 6 = ___

c) 8 – 3 = ___

d) 13 – 6 = ___

e)
$$\begin{array}{r} 6 \\ -\ 2 \\ \hline 4 \end{array}$$
/ / | | | |
1 2 3 4

f)
$$\begin{array}{r} 7 \\ -\ 4 \\ \hline \square \end{array}$$

g)
$$\begin{array}{r} 8 \\ -\ 2 \\ \hline \square \end{array}$$

h)
$$\begin{array}{r} 15 \\ -\ 3 \\ \hline \square \end{array}$$

2. Subtract by regrouping.

a) 64 – 31
= 6 tens + 4 ones – 3 tens + 1 one
= 6 – 3 tens + 4 – 1 ones
= 3 tens + 3 ones = 33

b) 79 – 57
= ___ tens + ___ ones – ___ tens + ___ ones
= ___ – ___ tens + ___ – ___ ones
= ______ tens + ______ ones = ______

c) 87 – 62
= ___ tens + ___ ones – ___ tens + ___ ones
= ___ – ___ tens + ___ – ___ ones
= ______ tens + ______ ones = ______

Worksheet-18

Subtraction Without and With Borrowing

1. Subtraction without borrowing.

a)

6	9
– 3	2

b)

7	4
– 4	1

c)

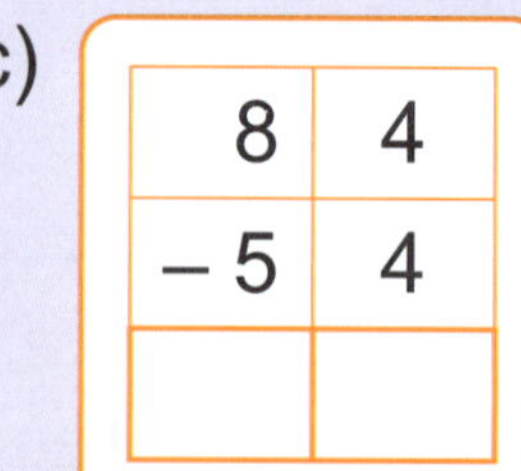

8	4
– 5	4

d)

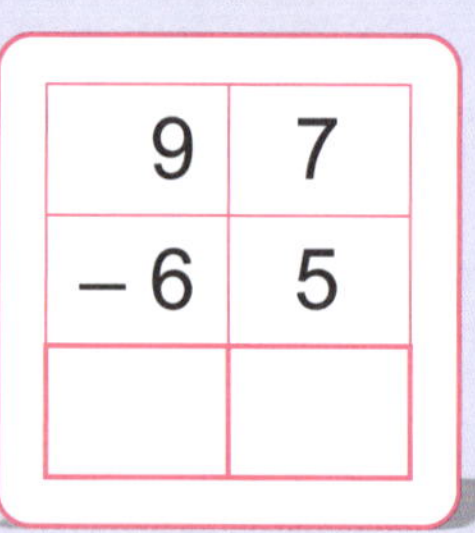

9	7
– 6	5

e) 85 – 33 = __________

f) 69 – 41 = __________

g) 57 – 25 = __________

h) 94 – 23 = __________

2. Subtraction with borrowing.

a)

6	4
– 2	7

b)

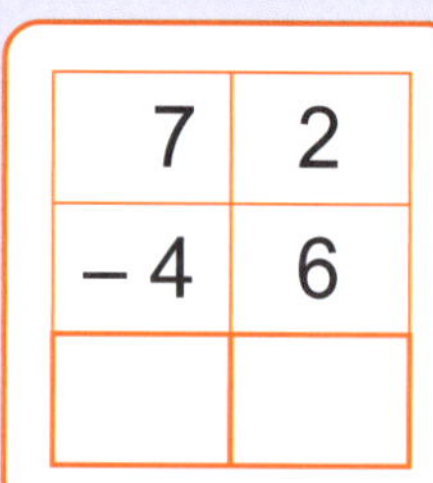

7	2
– 4	6

c)

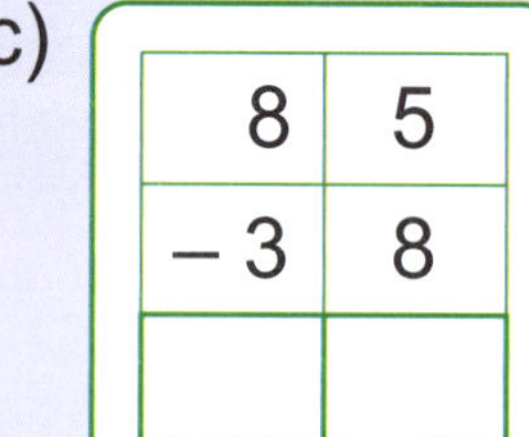

8	5
– 3	8

d)

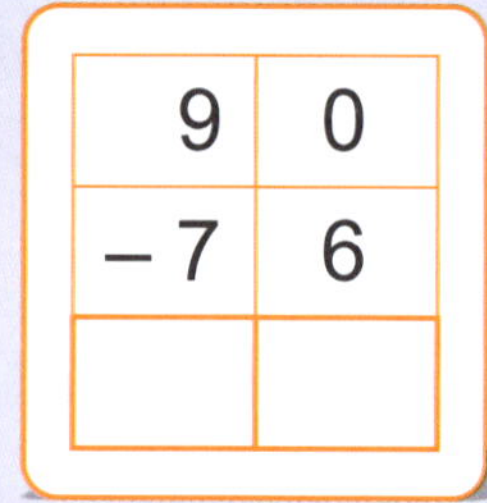

9	0
– 7	6

e) 83 – 26 = __________

f) 90 – 45 = __________

g) 73 – 28 = __________

h) 86 – 49 = __________

3. Simplify the following:

a) 66 – 43 + 34

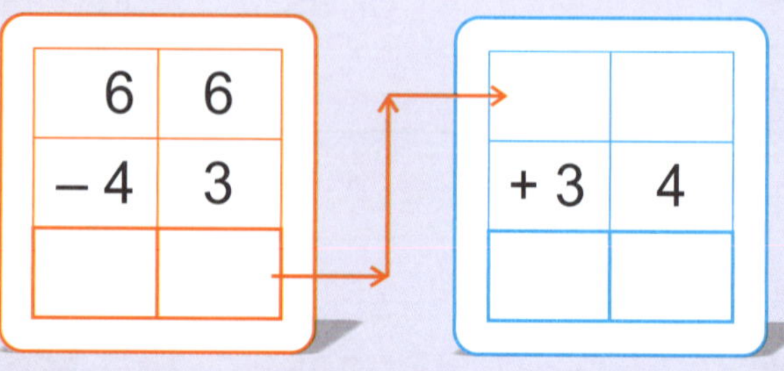

6	6
– 4	3

→

+ 3	4

b) 59 + 24 – 38

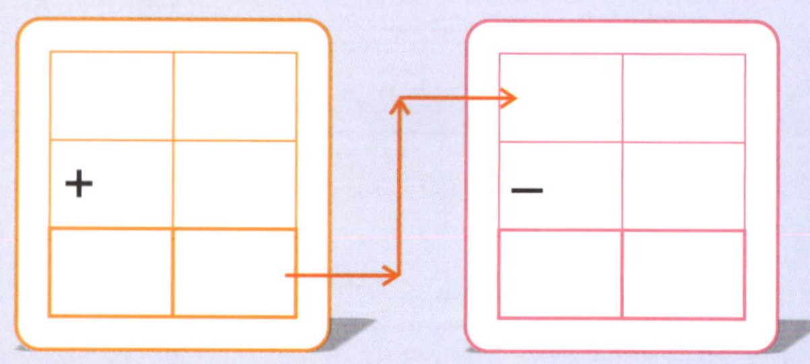

+	

→

–	

c) 63 + 26 – 35 = ________

d) 76 – 33 + 22 = ________

Worksheet-19

Subtraction Stories and Subtraction Facts

1. There were 55 passengers travelling in a chartered bus. At a bus stop 23 get off and nobody gets in. How many passengers are in the bus now?

____

____

____

2. Amy had 78 toys. She donated 15 of them. How many toys are left with her now?

____

____

____

3. A bookseller had 92 copies of a book. He sold 67 of them. How many are left with him now?

____

____

____

4. A jar had 48 candies. 26 of them were sold. Now 25 more were put in it. How many candies are in the jar now?

____

____

____

5. Fill in the blanks.

a) ______ – 45 = 0 b) 61 – ______ = 0 c) 92 – 92 = ______

d) ______ – 0 = 76 e) 31 – ______ = 31 f) 73 – 0 = ______

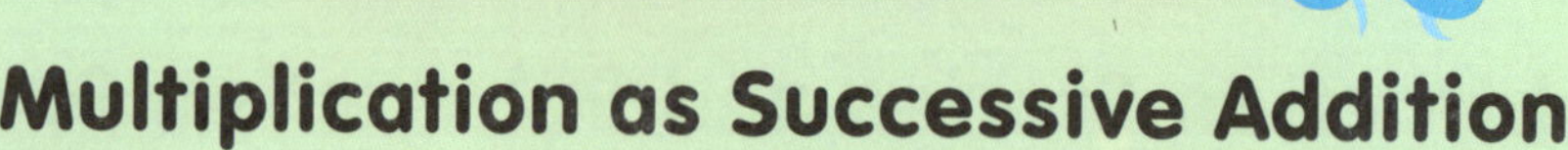

Worksheet-20

Multiplication as Successive Addition

1. Fill in the blanks.

a 

There are 2 bunches of balloons.
Each bunch has 3 balloons.
In all there are 3 + 3 = _____ balloons.
Or 2 times 3 is _____.
2 × 3 = _____.

b 

There are _____ baskets of toffees.
Each basket has _____ toffees.
In all there are ___+___+___+___=___ toffees.
Or _____ times _____ is _____.
_____×_____=_____.

c 

There are _____ jars of strawberries.
Each jar has _____ strawberries.
In all there are ___+___+___+___+___=___ strawberries.
Or _____ times _____ is _____.
_____×_____=_____.

2. Fill in the blanks.

a) 1 + 1 + 1 + 1 + 1 + 1 + 1 = _____ × 1

b) 2 + 2 + 2 + 2 + 2 + 2 + 2 + 2 = _____ × 2

c) 3 + 3 + 3 + 3 + 3 + 3 = 6 × _____

d) 4 +4 + 4 + 4 + 4 + 4 + 4 + 4 + 4 = 9 × _____

e) 7 × 5 = _____ + _____ + _____ + _____ + _____ + _____ + _____

f) 5 × 1 = _____ + _____ + _____ + _____ + _____

g) 2 + 2 + 2 + 2 + 2 + 2 + 2 + 2 = _____ × _____

h) 3 +3 + 3 + 3 + 3 + 3 + 3 + 3+ 3 = _____ × _____

Worksheet-21

Multiplication Using Number Line

Multiply by using number line.

1 Six times one: 6 × 1

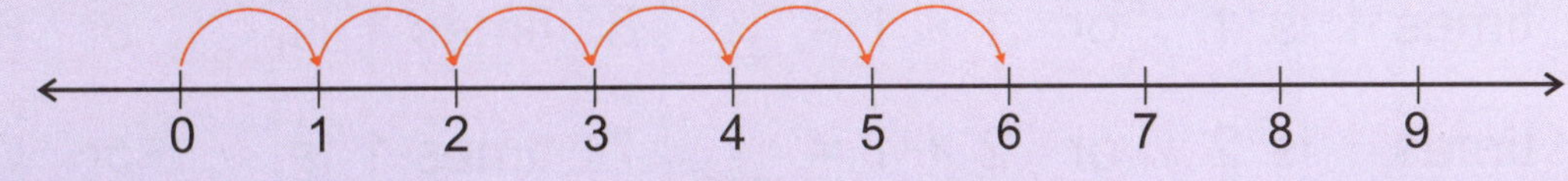

We start from 0 and take 6 jumps of 1 each. 6 × 1 = ☐

2 Four times two: 4 × 2

0 1 2 3 4 5 6 7 8 9

We start from ____ and take ____ jumps of ____ each. 4 × 2 = ☐

3 Five times three: 5 × 3

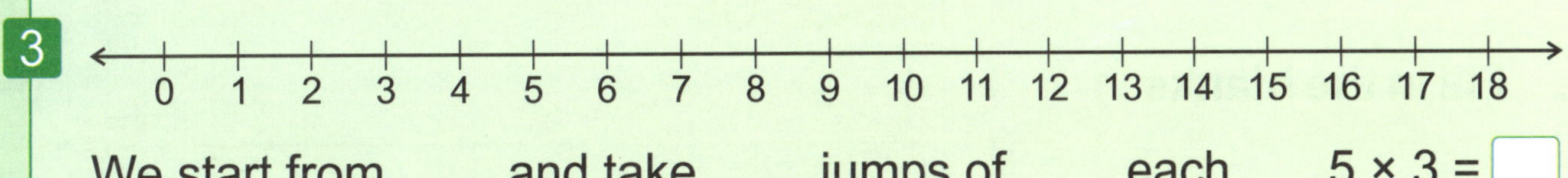

We start from ____ and take ____ jumps of ____ each. 5 × 3 = ☐

4 Seven times four: 7 × 4

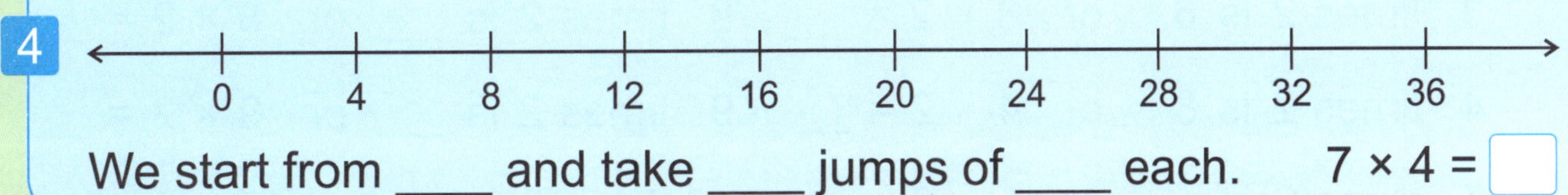

We start from ____ and take ____ jumps of ____ each. 7 × 4 = ☐

5 Eight times five: 8 × 5

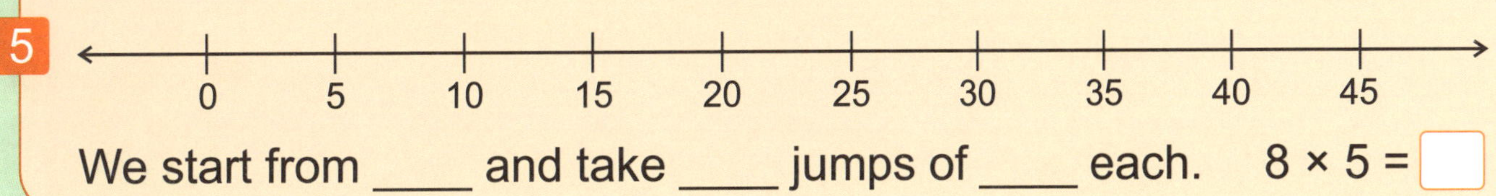

We start from ____ and take ____ jumps of ____ each. 8 × 5 = ☐

Worksheet-22

Tables of 1 and 2

TABLE OF 1

1. Fill in the blanks.

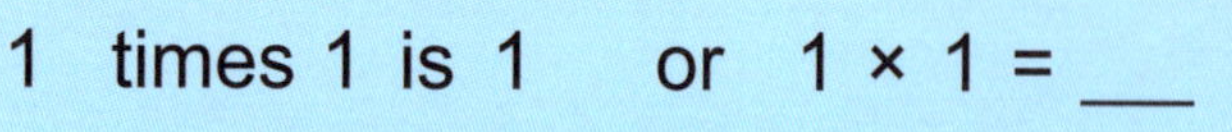

1 times 1 is 1 or 1 × 1 = ___	6 times 1 is ___ or 6 × 1 = ___
2 times 1 is 2 or 2 × 1 = ___	7 times 1 is ___ or 7 × 1 = ___
3 times 1 is 3 or 3 × 1 = ___	8 times 1 is ___ or 8 × 1 = ___
4 times 1 is 4 or 4 × 1 = ___	9 times 1 is ___ or 9 × 1 = ___
5 times 1 is 5 or 5 × 1 = ___	10 times 1 is ___ or 10 × 1 = ___

2. Fill in the boxes.

1	2								10

TABLE OF 2

3. Fill in the blanks.

1 times 2 is 2 or 1 × 2 = ___	6 times 2 is ___ or 6 × 2 = ___
2 times 2 is 4 or 2 × 2 = ___	7 times 2 is ___ or 7 × 2 = ___
3 times 2 is 6 or 3 × 2 = ___	8 times 2 is ___ or 8 × 2 = ___
4 times 2 is 8 or 4 × 2 = ___	9 times 2 is ___ or 9 × 2 = ___
5 times 2 is ___ or 5 × 2 = ___	10 times 2 is ___ or 10 × 2 = ___

4. Fill in the boxes.

2	4								20

Worksheet-23

Tables of 3 and 4

TABLE OF 3

1. Fill in the blanks.

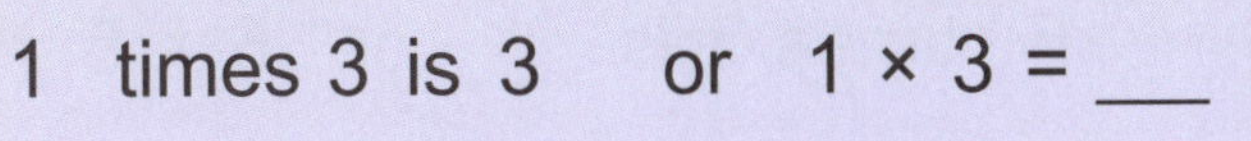

1 times 3 is 3 or 1 × 3 = ___	6 times 3 is ___ or 6 × 3 = ___
2 times 3 is 6 or 2 × 3 = ___	7 times 3 is ___ or 7 × 3 = ___
3 times 3 is 9 or 3 × 3 = ___	8 times 3 is ___ or 8 × 3 = ___
4 times 3 is ___ or 4 × 3 = ___	9 times 3 is ___ or 9 × 3 = ___
5 times 3 is ___ or 5 × 3 = ___	10 times 3 is ___ or 10 × 3 = ___

2. Fill in the boxes.

3	6								30

TABLE OF 4

3. Fill in the blanks.

1 times 4 is 4 or 1 × 4 = ___	6 times 4 is ___ or 6 × 4 = ___
2 times 4 is 8 or 2 × 4 = ___	7 times 4 is ___ or 7 × 4 = ___
3 times 4 is ___ or 3 × 4 = ___	8 times 4 is ___ or 8 × 4 = ___
4 times 4 is ___ or 4 × 4 = ___	9 times 4 is ___ or 9 × 4 = ___
5 times 4 is ___ or 5 × 4 = ___	10 times 4 is ___ or 10 × 4 = ___

4. Fill in the boxes.

4	8								40

Worksheet-24

Table of 5 and Dodging Tables

TABLE OF 5

1. Fill in the blanks.

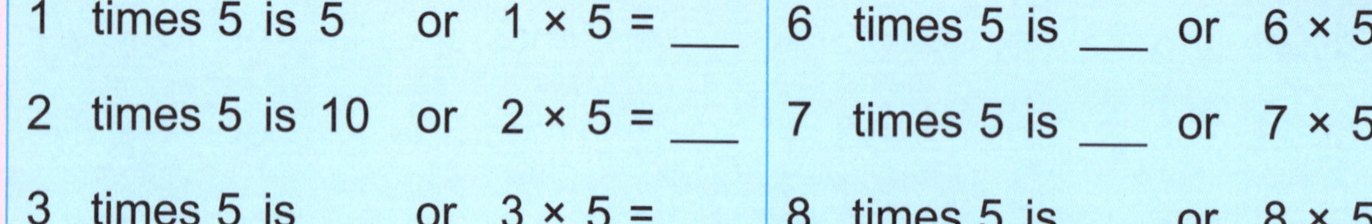

1 times 5 is 5 or 1 × 5 = ___	6 times 5 is ___ or 6 × 5 = ___
2 times 5 is 10 or 2 × 5 = ___	7 times 5 is ___ or 7 × 5 = ___
3 times 5 is ___ or 3 × 5 = ___	8 times 5 is ___ or 8 × 5 = ___
4 times 5 is ___ or 4 × 5 = ___	9 times 5 is ___ or 9 × 5 = ___
5 times 5 is ___ or 5 × 5 = ___	10 times 5 is ___ or 10 × 5 = ___

2. Fill in the boxes.

5	10								50

Dodging Tables

3. Fill in the blanks.

a)	6 × 4 =		b)	1 × 2 =		c)	6 × 1 =	
d)	8 × 1 =		e)	3 × 2 =		f)	5 × 4 =	
g)	5 × 5 =		h)	9 × 2 =		i)	10 × 4 =	
j)	4 × 3 =		k)	7 × 4 =		l)	8 × 2 =	
m)	7 × 2 =		n)	9 × 5 =		o)	4 × 1 =	
p)	9 × 3 =		q)	8 × 5 =		r)	9 × 4 =	

4. Multiply by 0.

a)	1 × 0 =		b)	2 × 0 =		c)	5 × 0 =	
d)	0 × 3 =		e)	0 × 4 =		f)	0 × 5 =	

Worksheet-25

Multiplication Without and With Carrying, Multiplication Facts

1. Multiplication without carrying.

a)

2	6
×	1

b)

3	4
×	2

c)

3	2
×	3

d)

2	1
×	4

e)

1	0
×	5

f) 68 × 1 = __________ g) 33 × 2 = __________

h) 33 × 3 = __________ i) 12 × 4 = __________

2. Multiplication with carrying.

a) 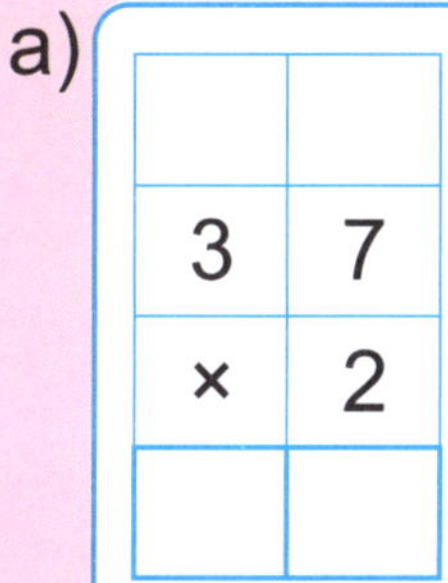

b) 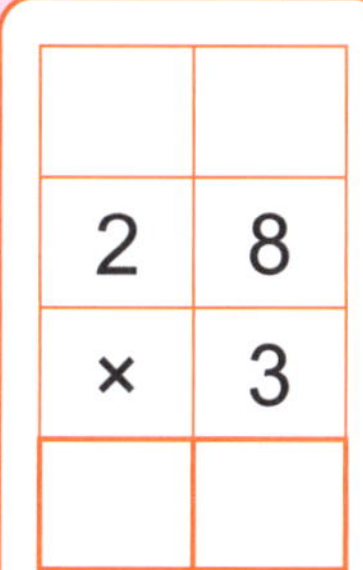

c) 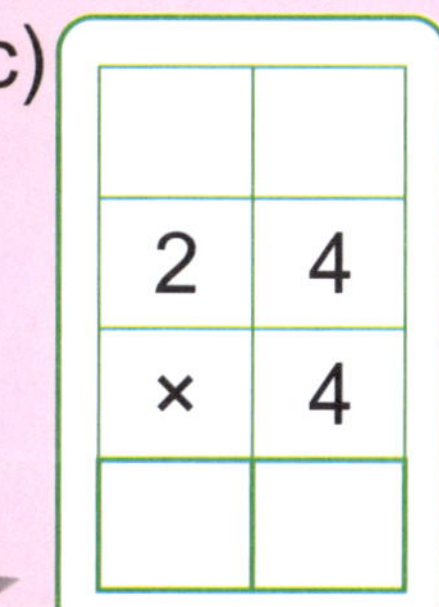

d) 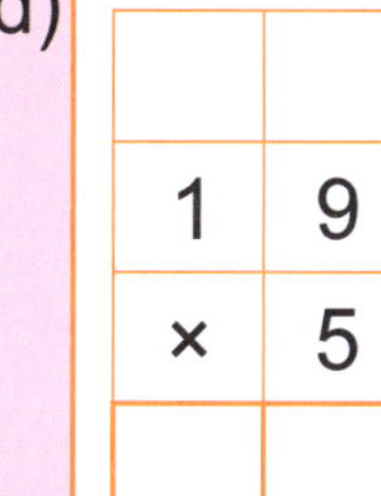

e) 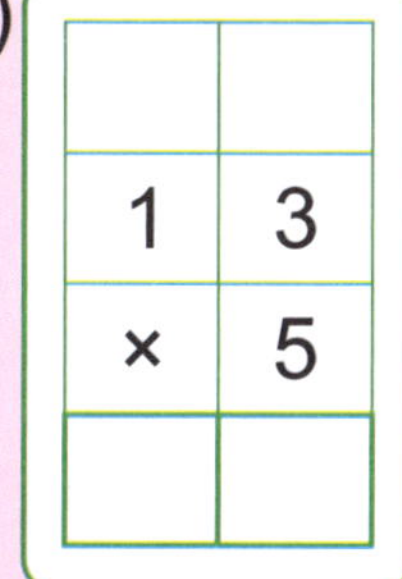

f) 48 × 2 = __________ g) 27 × 3 = __________

h) 13 × 4 = __________ i) 19 × 5 = __________

3. Fill in the blanks.

a) 3 × 1 = __________ b) 5 × __________ = 5

c) __________ × 1 = 4 d) 5 × 3 = __________ × 5

e) 3 × __________ = 7 × 3 f) __________ × 5 = 5 × 8

g) 3 × 0 = __________ h) 5 × __________ = 0

i) __________ × 2 = 0 j) 0 × 0 = __________

Worksheet-26

Multiplication Stories

1. A pouch has 32 erasers. How many erasers will 2 such pouches have?

2. A packet has 24 pencils. How many pencils will 3 such packets have?

3. A packet has 16 cans of juice. How many cans will 4 such packets have?

4. A mini bus can carry 16 passengers. How many passengers will 5 such buses carry?

5. Find sum of 24 and 18. Now multiply it by 2. What do you get?

Worksheet-27

Division as Equal Sharing and Equal Grouping

1. Division as equal sharing.

A boy has 9 toffees and 3 jars. He wants to put equal number of toffees in each jar.	
He puts 1 toffee in each jar. He has _____ toffees left.	
He puts 1 more toffee in each jar. He has _____ toffees left.	
He puts 1 more toffee in each jar. He has _____ toffees left.	

_______ toffees have been equally divided in _______ jars.

Each jar has _______ toffees.

_______ ÷ _______ = _______

2. Division as equal grouping.

a) There are 8 pencils.

They are in _____ equal groups.

There are _____ pencils in each group.

We say _____ divided by _____ is equal to _____.

We write _____ ÷ _____ = _____.

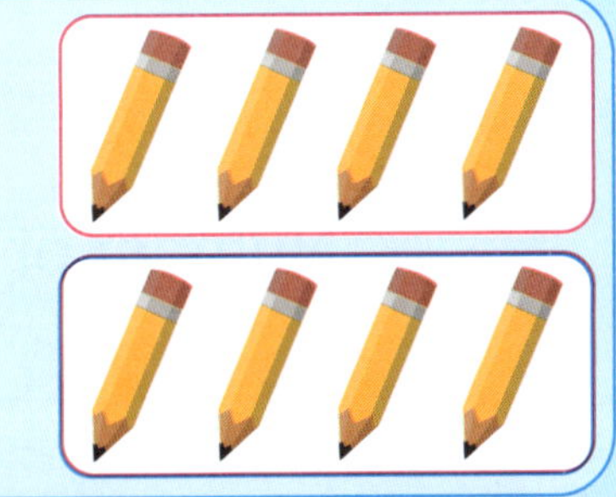

b) There are 12 apples.

They are in _____ equal groups.

There are _____ apples in each group.

We say _____ divided by _____ is equal to _____.

We write _____ ÷ _____ = _____.

Worksheet-28

Division as Successive Subtraction and Division Properties

1. Divide using successive subtraction.

a) 12 ÷ 2

Step 1: 12 – 2 = 10
Step 2: 10 – 2 = 8
Step 3: 8 – 2 = 6
Step 4: 6 – 2 = 4
Step 5: 4 – 2 = 2
Step 6: 2 – 2 = 0

Number of steps = 6

So 12 ÷ 2 = 6

b) 15 ÷ 3

____
____
____
____
____

Number of steps = ________

So ______________________

c) 16 ÷ 4

____
____
____
____
____

Number of steps = ________

So ______________________

d) 15 ÷ 5

____
____
____
____
____

Number of steps = ________

So ______________________

2. Fill in the blanks.

a) 2 ÷ 2 = ______
b) 3 ÷ ______ = 1
c) ______ ÷ 4 = 1
d) 5 ÷ 1 = ______
e) 2 ÷ ______ = 2
f) ______ ÷ 1 = 3

Worksheet-29

Long Division Without and With Remainder

1. Division without remainder.

a	b	c	d
$2\overline{)36}$ = 18 − 2 16 − 16 0 Quotient is 18.	$3\overline{)69}$ Quotient is ____.	$4\overline{)96}$ Quotient is ____.	$5\overline{)85}$ Quotient is ____.
e	**f**	**g**	**h**
$2\overline{)76}$ Quotient is ____.	$3\overline{)81}$ Quotient is ____.	$4\overline{)68}$ Quotient is ____.	$5\overline{)95}$ Quotient is ____.

2. Division with remainder.

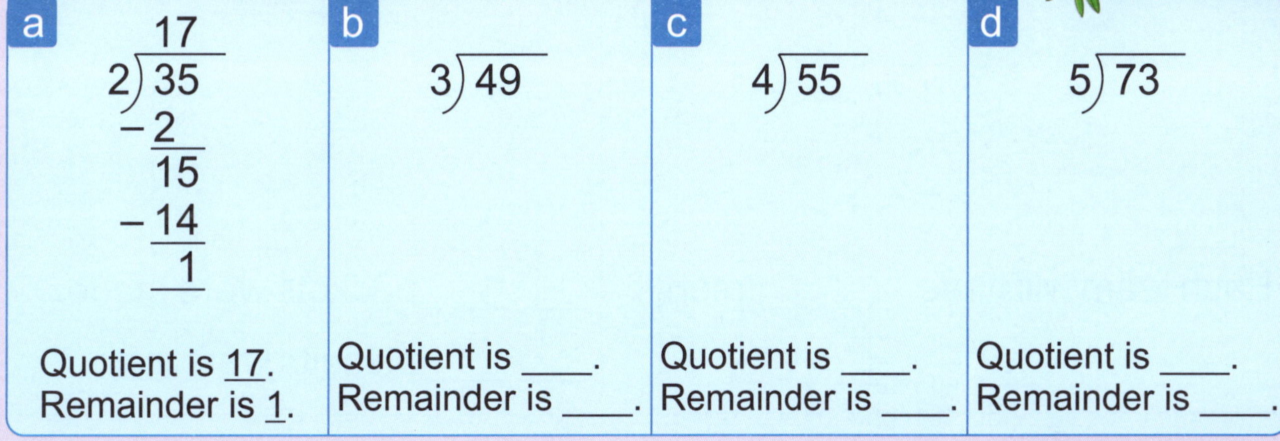

a	b	c	d
$2\overline{)35}$ = 17 − 2 15 − 14 1 Quotient is 17. Remainder is 1.	$3\overline{)49}$ Quotient is ____. Remainder is ____.	$4\overline{)55}$ Quotient is ____. Remainder is ____.	$5\overline{)73}$ Quotient is ____. Remainder is ____.

Worksheet-30

Division Stories

1. A carton has 84 toys. They have to be packed into packets of 2 toys each. How many packets can be made?

_______ packets can be made.

2. There are 66 eggs. They have to be packed into trays of 3 eggs each. How many trays can be made?

_______ trays can be made.

3. There are 48 men. They were divided into 4 equal teams. How many men will each team have?

Each team will have _______ men.

4. 69 biscuits were packed into packets of 5 each. How many packets were made? How many biscuits were left behind?

_______ packets were made.

_______ biscuits were left.

Worksheet-31

Increasing, Decreasing Order of Length

1. **Arrange in increasing order of length. Write 1, 2, 3 and 4 in the boxes.**

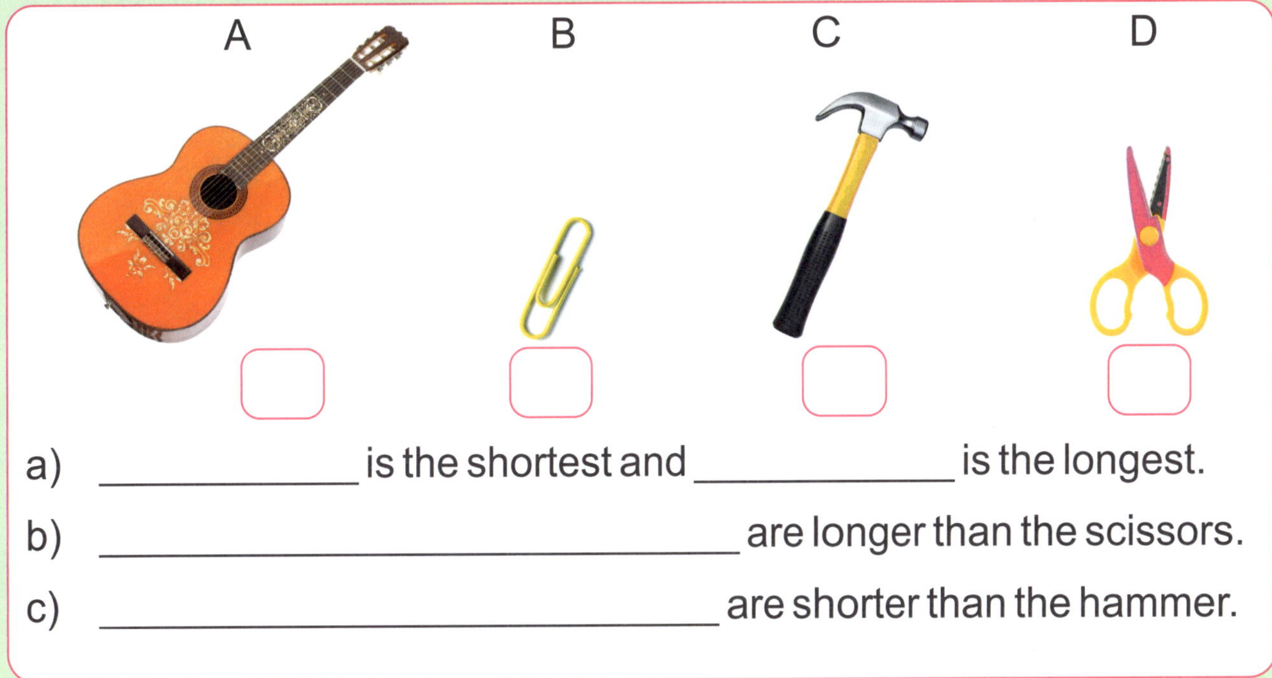

a) ____________ is the shortest and ____________ is the longest.

b) ______________________________ are longer than the scissors.

c) ______________________________ are shorter than the hammer.

2. **Arrange in decreasing order of height. Write 1, 2, 3 and 4 in the boxes.**

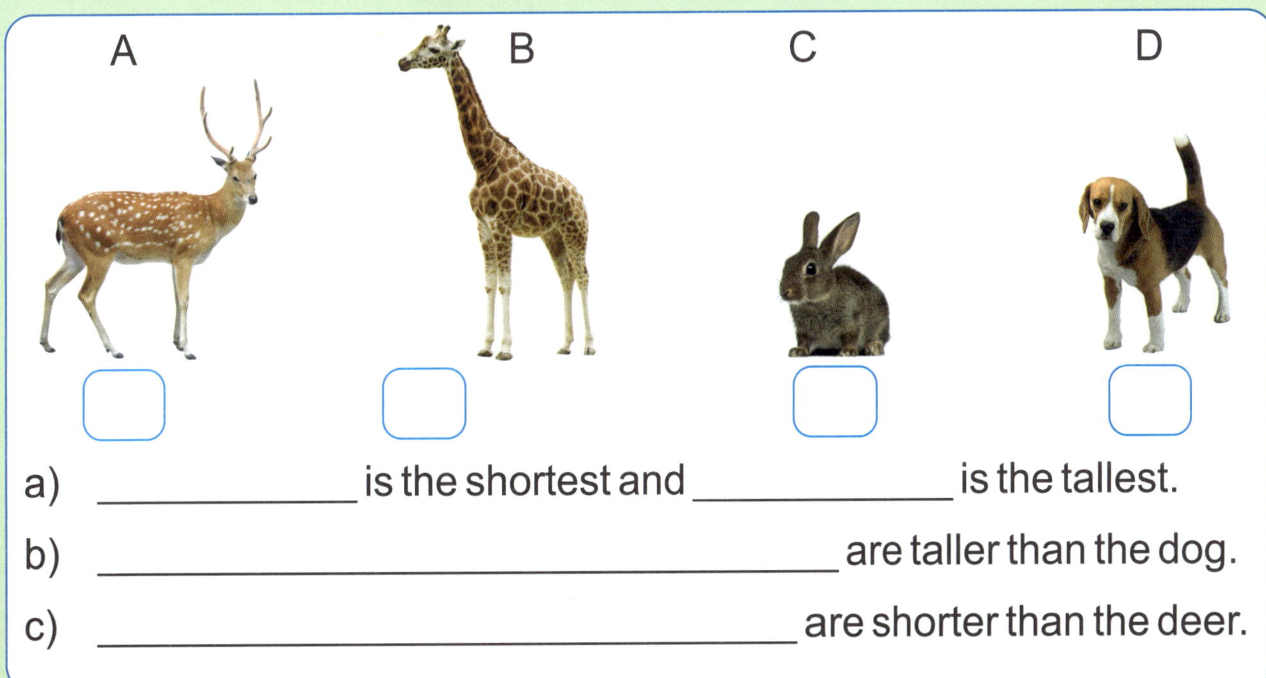

a) ____________ is the shortest and ____________ is the tallest.

b) ______________________________ are taller than the dog.

c) ______________________________ are shorter than the deer.

Worksheet-32

Measuring Length

1. Match the following:

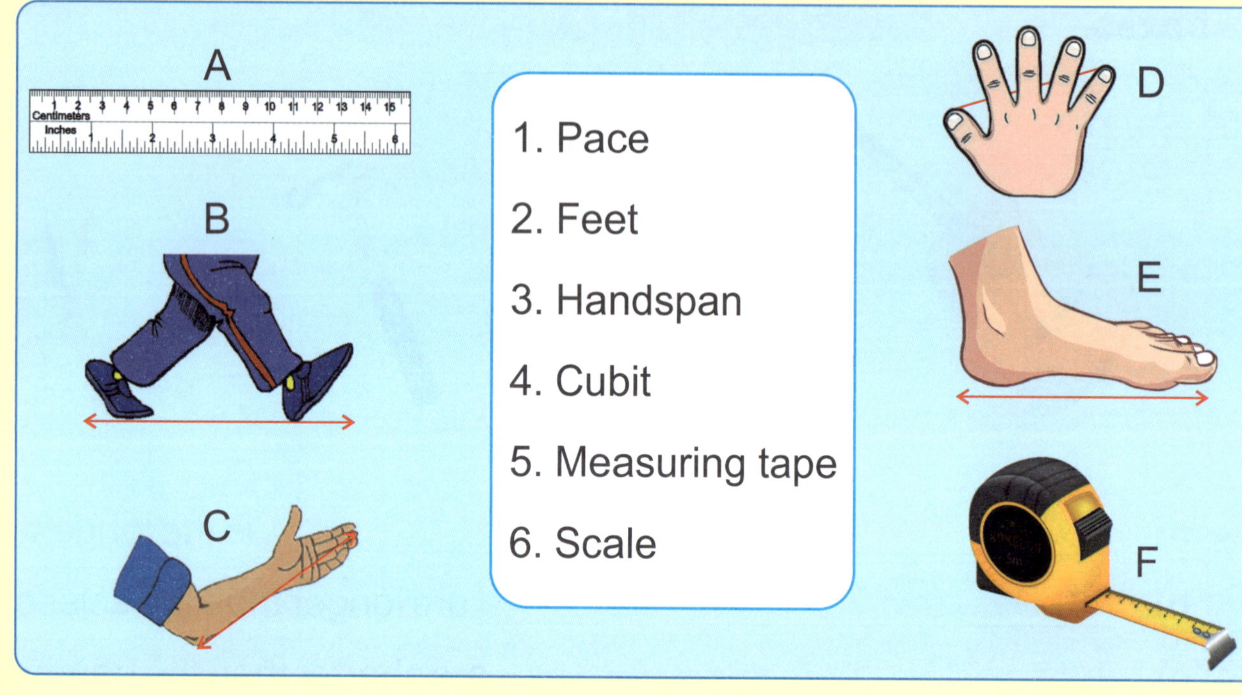

2. Fill in the blanks.

a) Non-standard units of length are cubit, ____________________.

b) Standard unit of length is ________________. Its symbol is _______.

c) To measure small lengths we use __________. Its symbol is _______.

d) ________________ cm = 1m.

e) A student uses ________________________ to measure length.

f) A mason uses ________________________ to measure length.

g) A cloth merchant uses ______________________ to measure length.

h) A tailor uses ________________________ to measure length.

i) A rope is 10 handspans long and a tie is 5 handspans long.
Which is longer? ________________ Which is shorter? ________

j) A rod is 5 cubits long. A wire is 5 handspans long.
Are they equal in length? ________________ (yes/no)

Worksheet-33

Increasing and Decreasing Order of Weight

1. **Arrange in increasing order of weight. Write 1, 2, 3 and 4 in the boxes.**

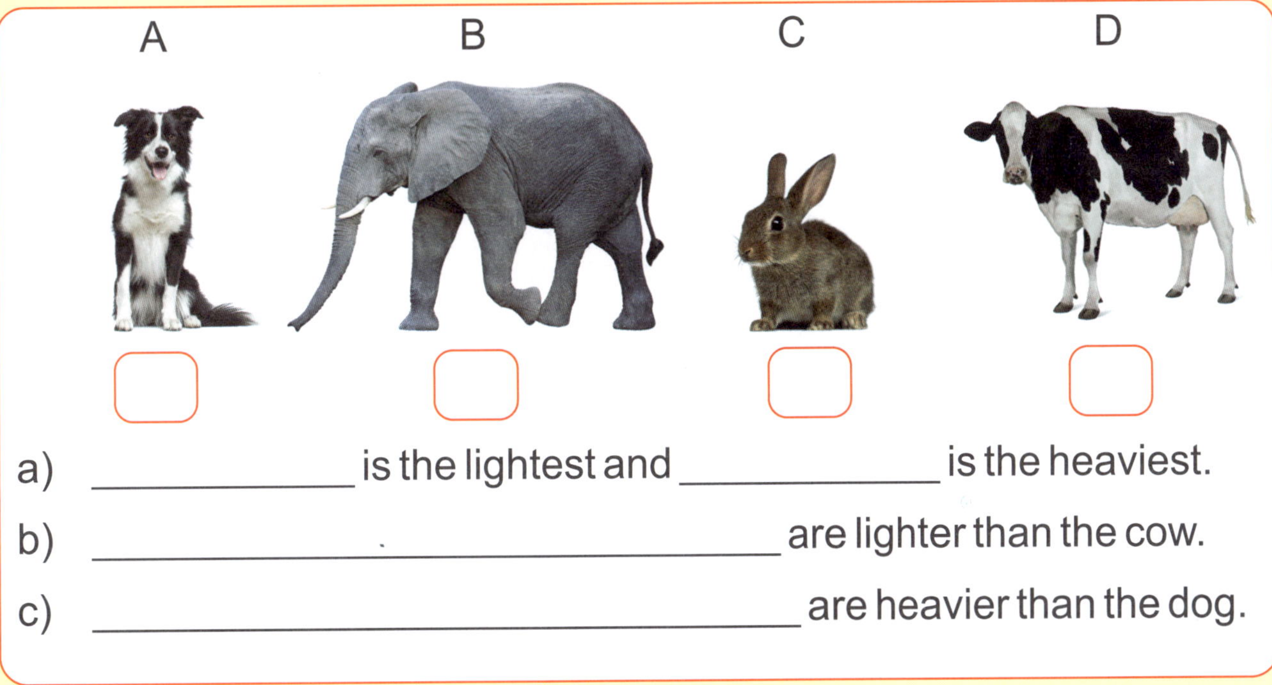

a) ____________ is the lightest and ____________ is the heaviest.

b) ________________________________ are lighter than the cow.

c) ________________________________ are heavier than the dog.

2. **Arrange in decreasing order of weight. Write 1, 2, 3 and 4 in the boxes.**

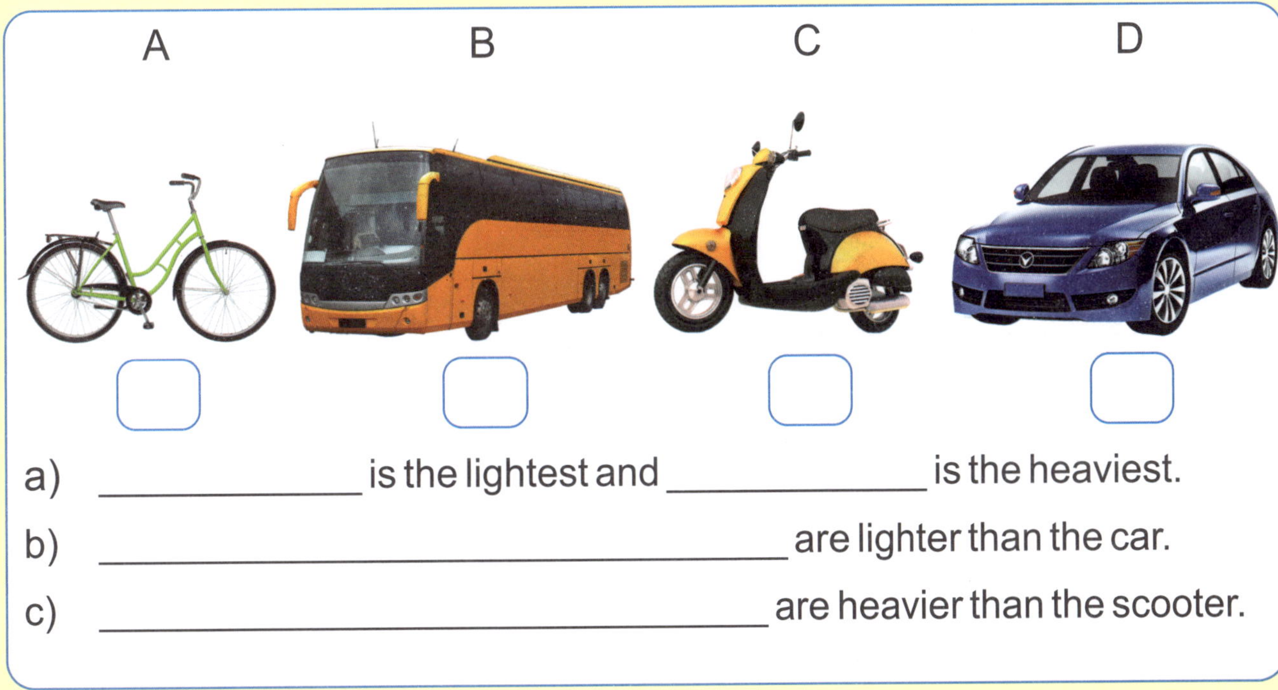

a) ____________ is the lightest and ____________ is the heaviest.

b) ________________________________ are lighter than the car.

c) ________________________________ are heavier than the scooter.

Worksheet-34

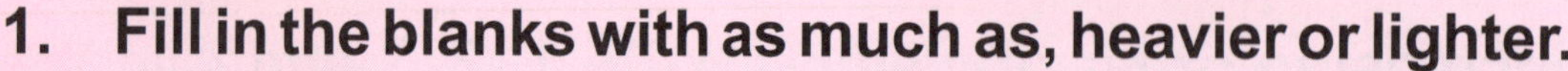

Comparing Weight

1. Fill in the blanks with as much as, heavier or lighter.

a  The leaves are ________________ than the pear.

b  The cups weigh ________________ the jar.

c  The pumpkin is ________________ than the carrot.

d  1 book weighs as much as ______ apples.

e ___________ bananas weigh as much as 1 pineapple.

2. Fill in the blanks.

a) Standard unit of weight is ________________. Its symbol is ______.

b) To measure a smaller weight we use ________. Its symbol is ______.

Worksheet-35

Increasing and Decreasing Order of Capacity

1. Arrange in increasing order of capacity. Write 1, 2, 3 and 4 in the boxes.

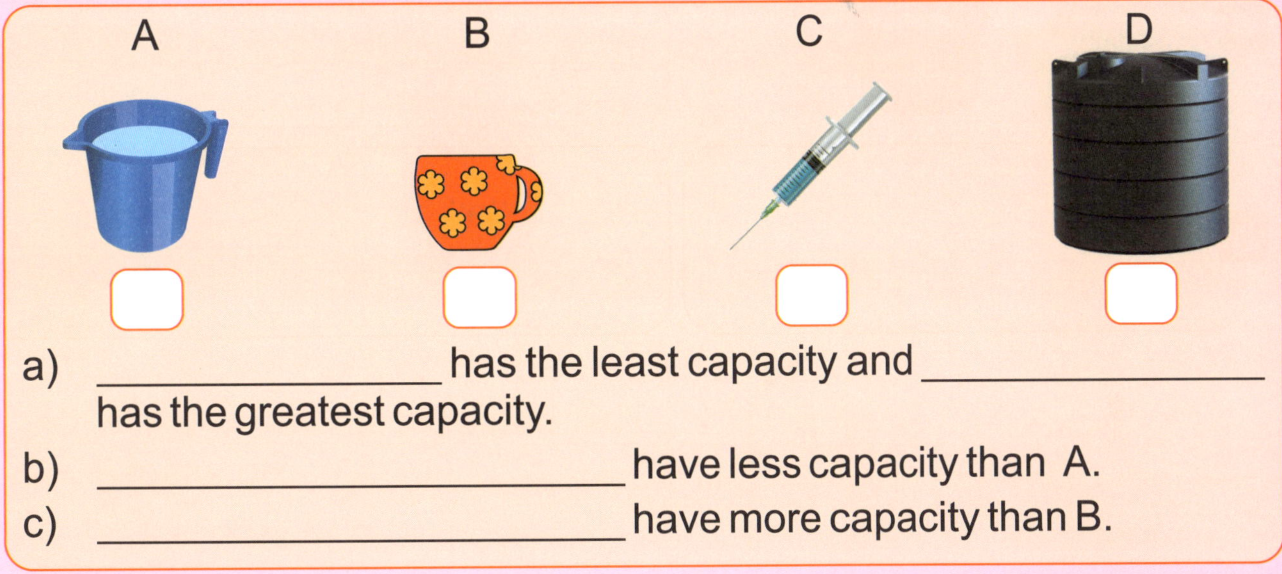

a) ________________ has the least capacity and ________________ has the greatest capacity.

b) __________________________ have less capacity than A.

c) __________________________ have more capacity than B.

2. Arrange in decreasing order of capacity. Write 1, 2, 3 and 4 in the boxes.

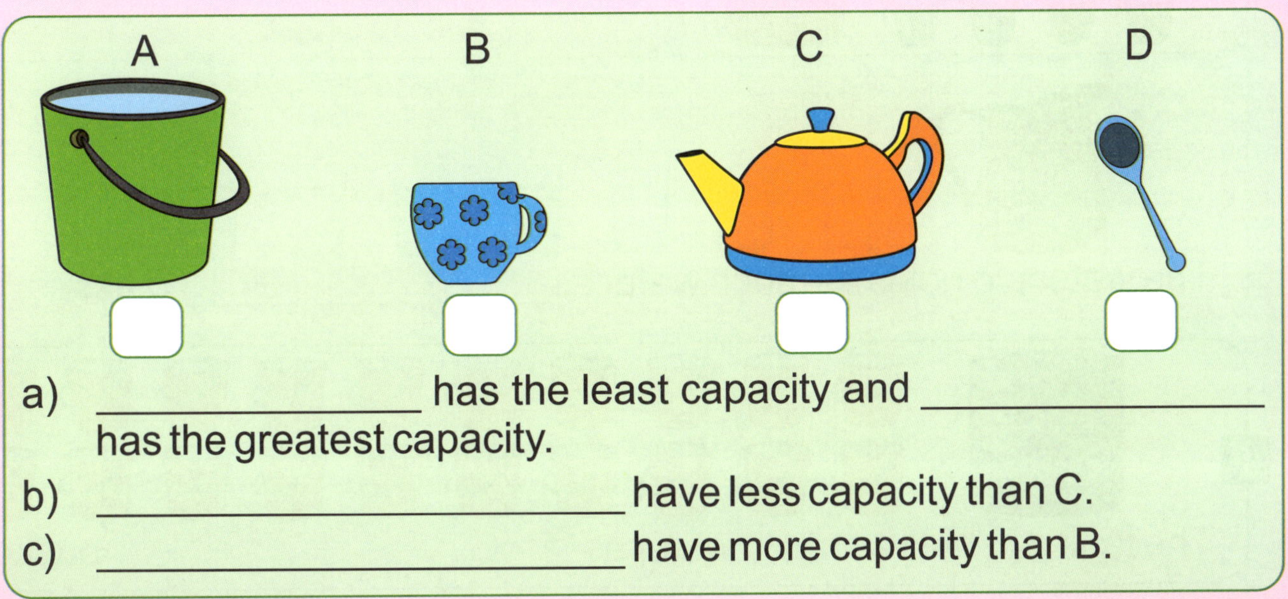

a) ________________ has the least capacity and ________________ has the greatest capacity.

b) __________________________ have less capacity than C.

c) __________________________ have more capacity than B.

3. Fill in the blanks.

a) The ____________ of a vessel tells us how much of water it can hold.

b) Standard unit of capacity is ___________. Its symbol is ___________.

c) To measure a smaller capacity we use __________. Its symbol is ___.

Worksheet-36

Comparing Capacity

Fill in the blanks.

a 

The bathroom mug has ______________ (more/less) capacity than the bucket.

b 

The bottle has ______________ (more/less) capacity than the glass.

c 

The bottle holds as much water as ______________ glasses.

d 

The bucket can hold as much water as ______________ bathroom mugs.

e 

The water jug holds as much water as ______________ glasses.

f 

The tank can hold as much water as ______________ buckets.

g  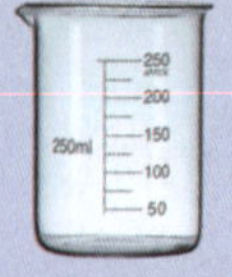 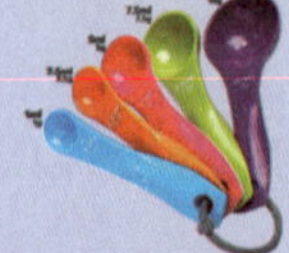 

______________ and ______________ are used to measure liquids in kitchen.

Worksheet-37

Time on Clocks

A clock has 2 hands: a long hand and a short hand.
The long hand tells us the minutes and the short hand tells us the hours.

1. What time is it?

2. Read the time below each clock and draw the hands.

6:15 8:20 6:40 1:50

2:15 3:45 8:15 9:25

Worksheet-38

Reading the Calendar

1

NOVEMBER

S	M	T	W	T	F	S
		1	2	3	4	5
6	7	8	9	10	11	12
13	14	15	16	17	18	19
20	21	22	23	24	25	26
27	28	29	30			

a) What month does the calendar show? ______________________

b) How many days are there in the month? ______________________

c) How many Fridays does the month have? ____________________

d) On what day of the week does the month start? _______________

e) On what date does the fourth Wednesday of the month falls on? _____

2

MARCH

S	M	T	W	T	F	S
31					1	2
3	4	5	6	7	8	9
10	11	12	13	14	15	16
17	18	19	20	21	22	23
24	25	26	27	28	29	30

a) If today is 12th, then what day of the month is it? _______________

b) On what day of the week does the month end? _________________

c) How many Sundays does the month have? ____________________

d) What date is the second Monday? __________________________

e) What date is the third Wednesday? _________________________

Worksheet-39

1. **Count and write the amount.**

Total amount
US$______

2. **Tick the smaller amount.**

A

B

3. **Tick the greater amount.**

A

B

4. **Tick the highest amount. Cross out the least amount.**

A

B

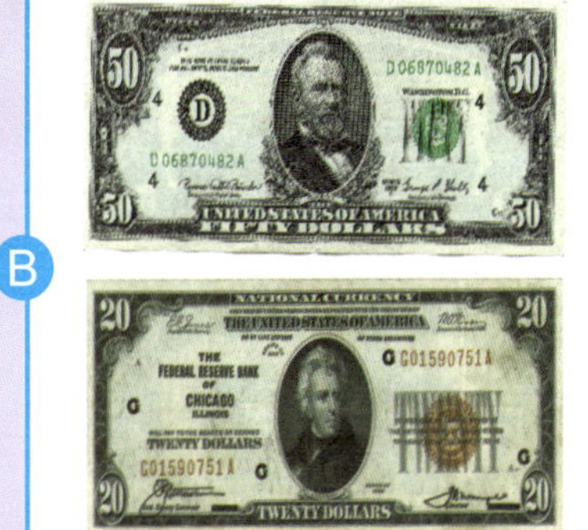

C

Worksheet-40

Buying and Selling

Rate card in a shop is as follows:

Item	Price
A	US$ 10
B	US$ 90
C	US$ 20

Item	Price
D	US$ 25
E	US$ 40
F	US$ 87

1. Answer the following:

a) Which items cost less than US$ 50? ____________________

b) Which item costs more than US$ 50 but less than US$ 90? ____________________

c) Which item is the costliest? ____________________

d) Which item is the cheapest? ____________________

e) If I buy items A and B, how much do I pay? US$ ____________________

2. Fill in the blanks.

a) Item ____________________ costs less than US$ 25 but is not the cheapest.

b) Items C and D together cost US$ ____________________.

c) I bought item E. I gave a note of US$ 50 to pay for it. I got back US$ ____________________.

d) I bought item ____________________. I gave a note of US$ 100 and got back a note of US$ 10.

e) I bought item D. I gave two notes of US$ ____________________ and one note of US$ 5 to pay for it.

Worksheet-41

Count and Write

John was asked to name his friends.

He wrote ten names.

1. Let us count the number of letters in each name.

SANDRA	Letters
JULIE	Letters
HELEN	Letters
CLARA	Letters
JOSEPH	Letters
MEGAN	Letters
ROBERT	Letters
MICHAEL	Letters
JENNY	Letters
JESSICA	Letters

2. a) How many names have 5 letters?

   b) How many names have 6 letters?

   c) How many names have 7 letters?

   d) How many times does 'A' occur?

   e) How many times does 'J' occur?

   f) How many times does 'L' occur?

Worksheet-42

Count and Write

Colour all △ as green.
Colour all ○ as blue.
Colour all □ as yellow.
Colour all ▭ as red.
Colour all ☆ as pink.
Colour all ⬭ as orange

1. How many □ ? ______________
2. How many ▭ ? ______________
3. How many ○ ? ______________
4. How many △ ? ______________
5. How many ☆ ? ______________
6. How many ⬭ ? ______________
7. Draw the shape that occurs most. ______
8. Draw the shape that occurs least. ______

Worksheet-43

Count and Colour

Count the vehicles and colour as many boxes. There are 5 scooters so 5 boxes have been coloured.

Worksheet-44

Completing Patterns

1. What comes next?

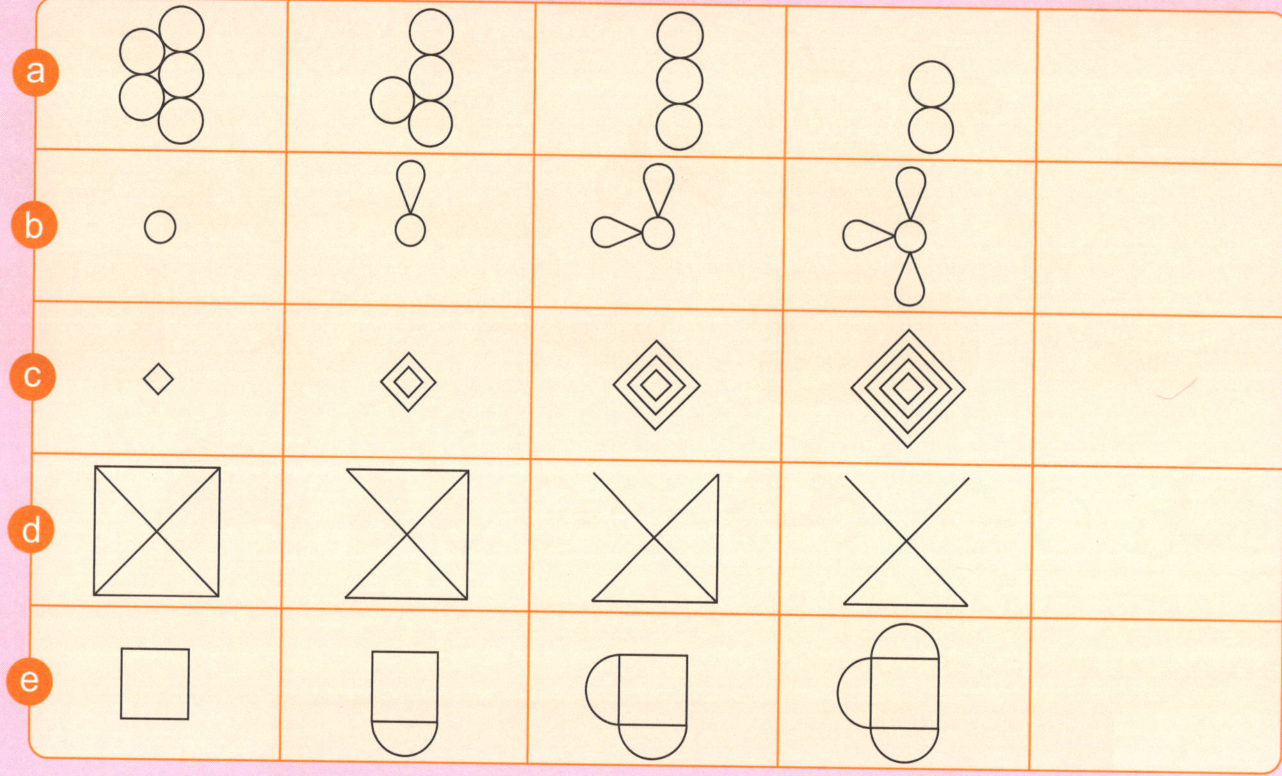

2. Extend the sequence.

Worksheet-45

Reducing and Growing Patterns

1. Complete the following reducing patterns:

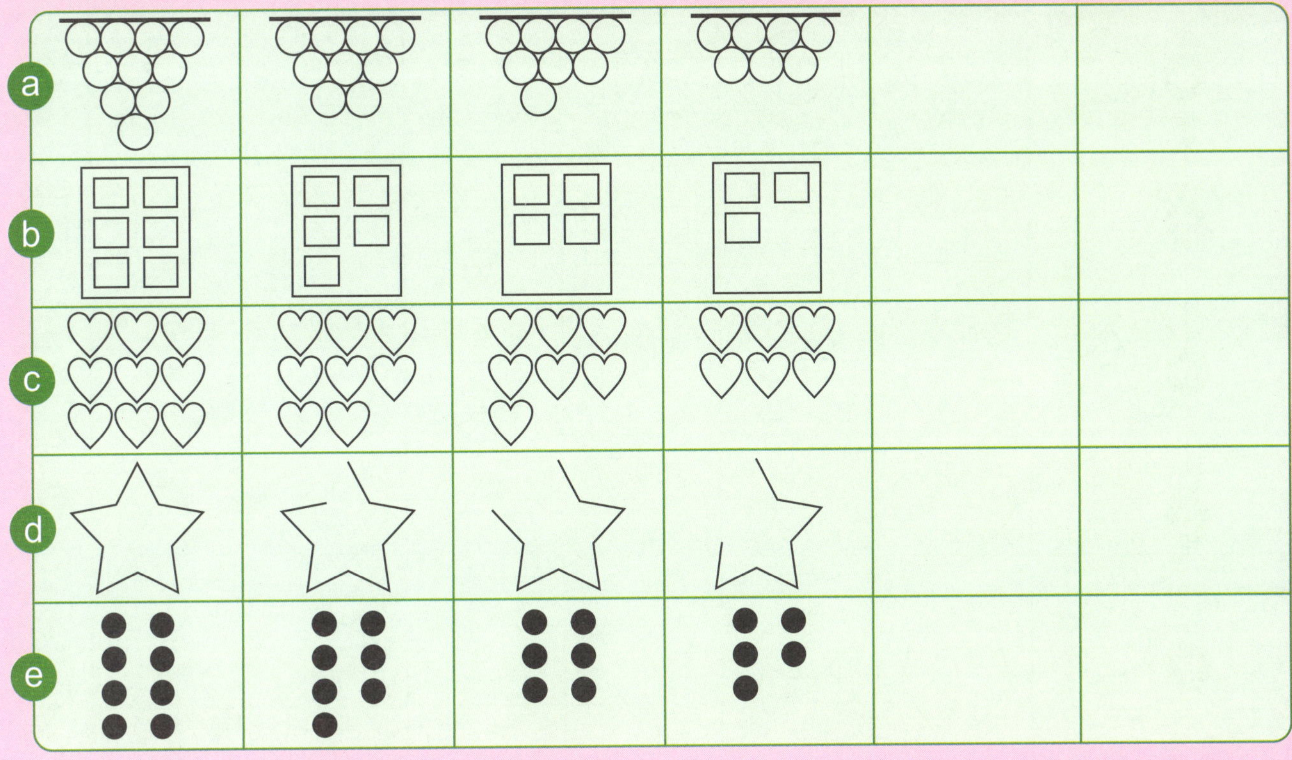

2. Complete the following growing patterns:

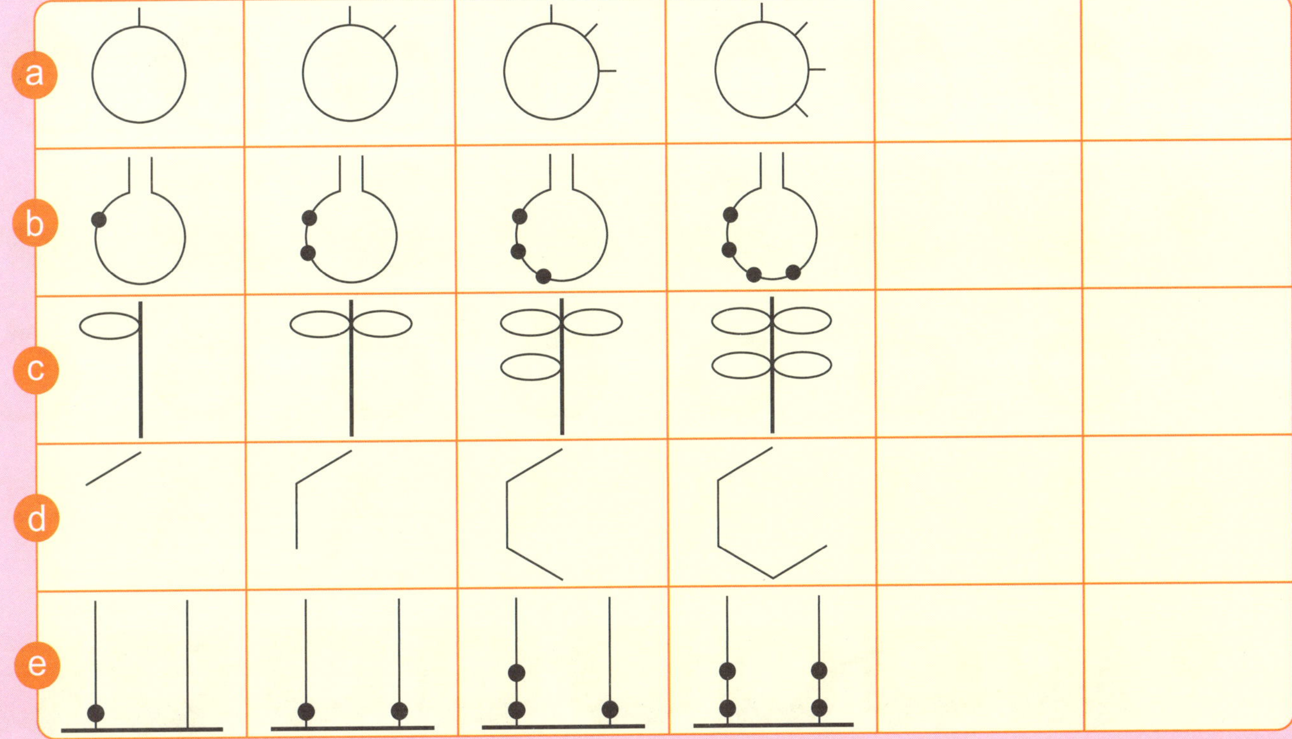

Worksheet-46

Filling Missing Patterns and Number Patterns

1. Fill the missing patterns in the empty boxes.

2. Fill in the missing numbers.

Worksheet-47

Finding Correct Shape-I

1. Tick the objects shaped like a triangle.

2. Tick the objects shaped like a square.

3. Tick the objects shaped like a rectangle.

4. Tick the objects shaped like a circle.

Worksheet-48

Finding Correct Shape-II

1. **Tick the objects shaped like a cube.**

A

B

C

D

E

2. **Tick the objects shaped like a cuboid.**

A

B

C

D

E

3. **Tick the objects shaped like a cylinder.**

A

B

C

D

E

4. **Tick the objects shaped like a cone.**

A

B

C

D

E

5. **Tick the objects shaped like a sphere.**

A

B

C

D

E

Worksheet-49

Sorting Shapes

Draw lines to join the same shapes.

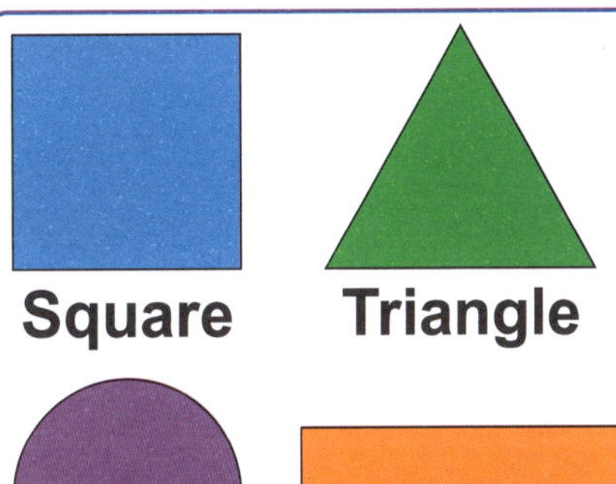
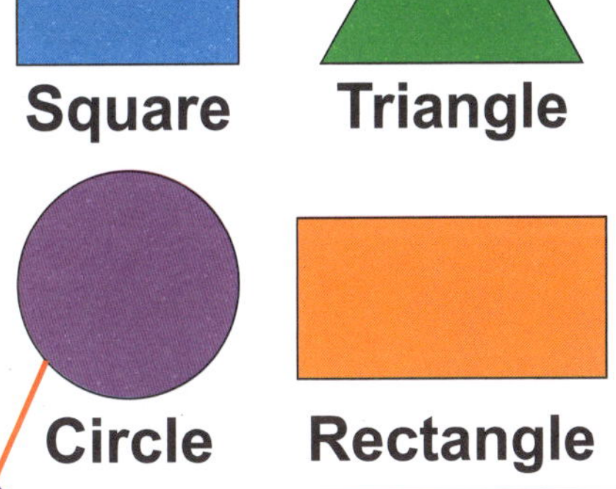

Worksheet-50

Sorting Shapes
Roll and Slide

1. **Draw lines to join the same shapes.**

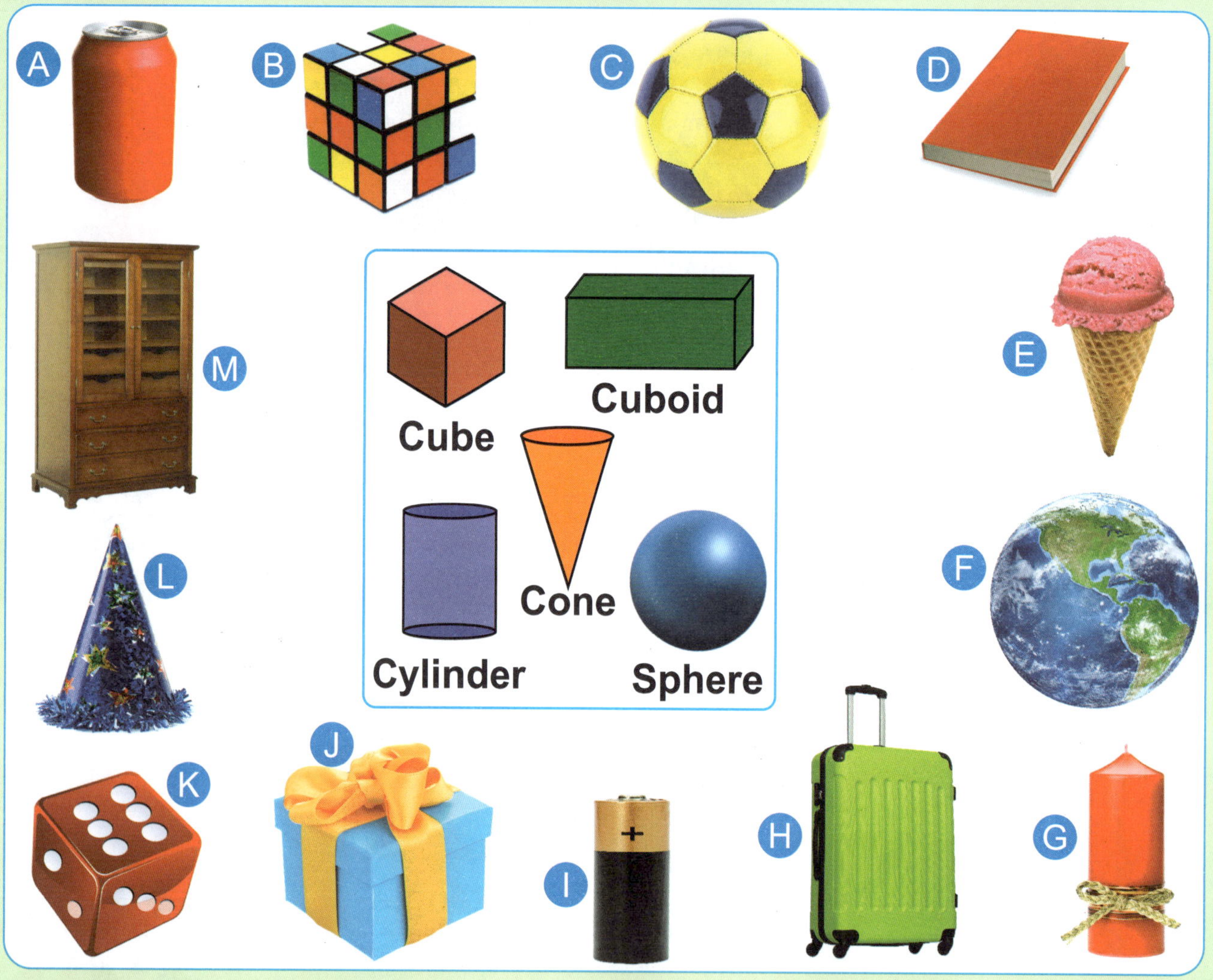

A B C D

M E

Cube

Cuboid

Cone

Cylinder

Sphere

L F

K J I H G

2. **Tick the objects that will roll when pushed. Cross the objects that will slide when pushed.**

ANSWERS

Unit-1 Numbers: 1 to 100

Worksheet-1

1. b) 3; three c) 6; six d) 8; eight e) 9; nine
2. b) 4; four c) 5; five d) 8; eight e) 7; seven

Worksheet-2

1. a) 3, 4, 5, 6, 7, 8 b) 18, 20, 21, 22, 23
   c) 30, 31, 33, 34, 35, 36 d) 61, 62, 63, 65, 66
   e) 93, 94, 95, 96, 97, 99
2. a) 6, 5, 4, 3, 2 b) 26, 25, 23, 22, 21
   c) 44, 43, 42, 41, 40, 39 d) 86, 85, 84, 83
   e) 98, 97, 96, 94, 93

Worksheet-3

8	Nine
13	Seventeen
25	Twenty-one
34	Thirty-seven
49	Forty-three
51	Fifty-five
58	Fifty-nine
68	Sixty-one
66	Sixty-four
70	Seventy-three
78	Seventy-six
84	Eighty-five
91	Ninety-four
100	Ninety-nine

Worksheet-4

1. a) 2, 3 b) 4, 9 c) 3, 6 d) 5, 0
2. a) 32 b) 54 c) 76 d) 91
3. a) twenty-seven b) fifty-eight
   c) eighty-four d) ninety-six
4. a) 4 b) 25 c) 74, 75 d) 97, 98, 99
5. a) 8 b) 33 c) 65, 66 d) 98, 99, 100
6. a) 8 b) 35 c) 76, 77 d) 97, 98, 99
7. a) 3, 5 b) 33, 34, 36, 37
   c) 94, 95, 96, 98, 99, 100

Worksheet-5

1. a) 2, 4 b) 6, 7
2. a) 40, 3 b) 80, 1
3. a) 70 + 3 b) 80 + 5
4. a) 58 b) 75 c) 48
5. a) < b) > c) = d) > e) < f) =
6. a) 67, 49, 14 b) 92, 56, 53, 48
7. a) 21, 35, 67, 92 b) 28, 31, 35, 53, 100

Worksheet-6

1.
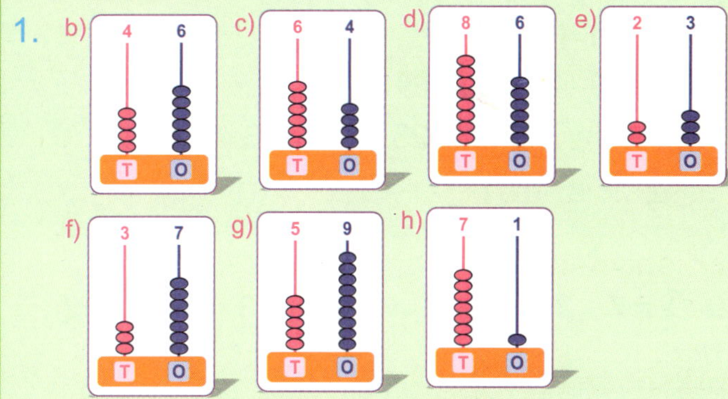

2. b) 55 c) 19 d) 28 e) 36 f) 43 g) 80 h) 99

Worksheet-7

1. 36; 63 2. 22; 88 3. 50; 50 4. 90; 99 5. 46; 96
6. 70; 99 7. 44, 49, 94, 99 44; 99
8. 11, 13, 15, 31, 33, 35, 51, 53, 55 11; 55

Worksheet-8

1.

Numeral	Number name	Ordinal Number	Also written as
1	One	First	1st
2	Two	Second	2nd
3	Three	Third	3rd
4	Four	Fourth	4th
5	Five	Fifth	5th
6	Six	Sixth	6th
7	Seven	Seventh	7th
8	Eight	Eighth	8th
9	Nine	Ninth	9th
10	Ten	Tenth	10th

2. a) J b) D c) third d) second e) right f) left
   g) sixth, fifth h) left, right

Unit-2 Operations on Numbers

Worksheet-9

2. 5 + 2 = 7 3. 4 + 4 = 8 4. 4 + 2 = 6 5. 3 + 5 = 8

Worksheet-10

1. b) 2 + 5 = 7 c) 5 + 8 = 13
2. b) 5 c) 6

Worksheet-11
1. b) 10 c) 12 d) 15 f) 7 g) 10 h) 14
2. b) 69 c) 89

Worksheet-12
1. a) 68 b) 89 c) 37 d) 97 e) 84 f) 87 g) 58 h) 49 i) 30 j) 68
2. a) 65 b) 41 c) 56 d) 74 e) 71 f) 88 g) 56 h) 65
3. a) 72 b) 97 c) 92 d) 85 e) 95 f) 90

Worksheet-13
1. 15 2. 71 3. 37 4. 94
5. a) 4 b) 3 c) 54 d) 23 e) 25, 43 f) 21, 44 g) 62 h) 0 i) 81

Worksheet-14
2. 9 – 2 = 7 3. 5 – 2 = 3 4. 6 – 3 = 3 5. 4 – 2 = 2

Worksheet-15
1. 3 2. 3 3. 2 4. 5 5. 5

Worksheet-16
1. b) 8 – 5 = 3 c) 11 – 4 = 7
2. b) 4 c) 8

Worksheet-17
1. b) 3 c) 5 d) 7 e) 4 f) 3 g) 6 h) 12
2. b) 22 c) 25

Worksheet-18
1. a) 37 b) 33 c) 30 d) 32 e) 52 f) 28 g) 32 h) 71
2. a) 37 b) 26 c) 47 d) 14 e) 57 f) 45 g) 45 h) 37
3. a) 57 b) 45 c) 54 d) 65

Worksheet-19
1. 32 2. 63 3. 25 4. 47
5. a) 45 b) 61 c) 0 d) 76 e) 0 f) 73

Worksheet-20
1. a) 6 b) 12 c) 20
2. a) 7 b) 8 c) 3 d) 4 e) 5 + 5 + 5 + 5 + 5 + 5 + 5 f) 1 + 1 + 1 + 1 + 1 g) 8 × 2 h) 9 × 3

Worksheet-21
1. 6 2. 8 3. 15 4. 28 5. 40

Worksheet-22
1.

1	6, 6
2	7, 7
3	8, 8
4	9, 9
5	10, 10

2.

1	2	3	4	5	6	7	8	9	10

3.

2	12, 12
4	14, 14
6	16, 16
8	18, 18
10, 10	20, 20

4.

2	4	6	8	10	12	14	16	18	20

Worksheet-23
1.

3	18, 18
6	21, 21
9	24, 24
12, 12	27, 27
15, 15	30, 30

2.

3	6	9	12	15	18	21	24	27	30

3.

4	24, 24
8	28, 28
12, 12	32, 32
16, 16	36, 36
20, 20	40, 40

4.

4	8	12	16	20	24	28	32	36	40

Worksheet-24
1.

5	30, 30
10	35, 35
15, 15	40, 40
20, 20	45, 45
25, 25	50, 50

2.

5	10	15	20	25	30	35	40	45	50

3. a) 24 b) 2 c) 6 d) 8 e) 6 f) 20 g) 25 h) 18 i) 40 j) 12 k) 28 l) 16 m) 14 n) 45 o) 4 p) 27 q) 40 r) 36
4. a) 0 b) 0 c) 0 d) 0 e) 0 f) 0

Worksheet-25

1. a) 26 b) 68 c) 96 d) 84 e) 50 f) 68 g) 66
   h) 99 i) 48
2. a) 74 b) 84 c) 96 d) 95 e) 65 f) 96 g) 81
   h) 52 i) 95
3. a) 3 b) 1 c) 4 d) 3 e) 7 f) 8 g) 0 h) 0
   i) 0 j) 0

Worksheet-26

1. 64 2. 72 3. 64 4. 80 5. 84

Worksheet-27

1. 3 2. a) 2 b) 4

Worksheet-28

1. b) 5 c) 4 d) 3
2. a) 1 b) 3 c) 4 d) 5 e) 1 f) 3

Worksheet-29

1. b) 23 c) 24 d) 17 e) 38 f) 27 g) 17 h) 19
2. b) 16; 1 c) 13; 3 d) 14; 3

Worksheet-30

1. 42 2. 22 3. 12 4. 13; 4

Unit-3 Measurement

Worksheet-31

1. A – 4, B – 1, C – 3, D – 2
   a) B; A d) A and C c) B and D
2. A – 2, B – 1, C – 4, D – 3
   a) C; B b) A and B c) C and D

Worksheet-32

1. A – 6, B – 1, C – 4, D – 3, E – 2, F – 5
2. a) handspan and pace b) metre; m
   c) centimetre; cm d) 100 e) scale
   f) measuring tape g) metre rod
   h) inch – tape i) rope; tie j) no

Worksheet-33

1. A – 2, B – 4, C – 1, D – 3
   a) C; B b) A and C c) B and D
2. A – 4, B – 1, C – 3, D – 2
   a) A; B b) A and C c) B and D

Worksheet-34

1. a) lighter b) as much as c) heavier
   d) 3 e) 6
2. a) kilogram; kg b) gram; g

Worksheet-35

1. A – 3, B – 2, C – 1, D – 4
   a) C; D b) B and C c) A and D
2. A – 1, B – 3, C – 2, D – 4
   a) D; A b) B and D c) A and C
3. a) capacity b) litre; l c) millilitre; ml

Worksheet-36

a) less b) more c) 4 d) 8 e) 8 f) 20
g) measuring jars; measuring spoons

Unit-4 Time

Worksheet-37

1. 11:45, 6:15, 3:20, 11:15, 7:25,
   8:30, 9:00, 10:35
2.

Worksheet-38

1. a) November b) 30 c) 4 d) Tuesday e) 23rd
2. a) Tuesday b) Sunday c) 5 d) 11th e) 20th

Unit-5 Money

Worksheet-39

1. Total amount = US$ 88
2. B) US$ 62 3. A) US$ 72
4. Box B has the highest amount (US$ 70).
   Box C has the least amount (US$ 20).

Worksheet-40

1. a) A, C, D and E b) F c) B d) A e) 100
2. a) C b) 45 c) 10 d) B e) 10

Unit-6 Data Handling

Worksheet-41

1. 6, 5, 5, 5, 6, 5, 6, 7, 5, 7

2. a) 5 b) 3 c) 2 d) 7 e) 4 f) 4

Worksheet-42

1. 9 2. 7 3. 3 4. 5 5. 6 6. 4 7. □ 8. ○

Worksheet-43

Unit-7 Patterns

Worksheet-44

Worksheet-45

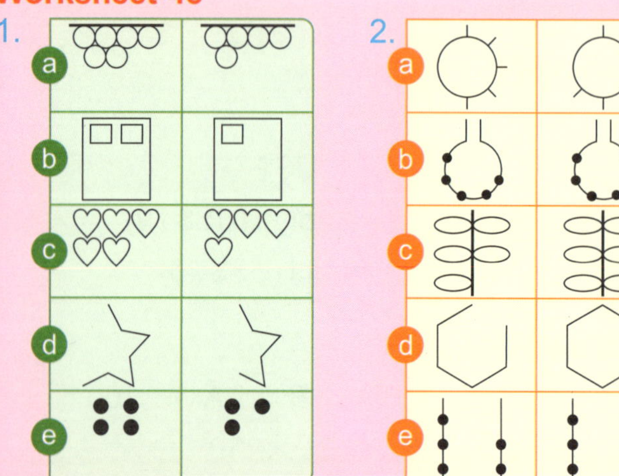

Worksheet-46

2. a) 30; 50; 70
   b) 35; 55; 65
   c) 33; 55
   d) 45; 67
   e) 15; 25

Unit-8 Shapes

Worksheet-47

1. B, C, E 2. B, D, E
3. B, C, E 4. A, B, D

Worksheet-48

1. C, E 2. A, B, E 3. A, C, E
4. A, D 5. A, D

Worksheet-49

Worksheet-50

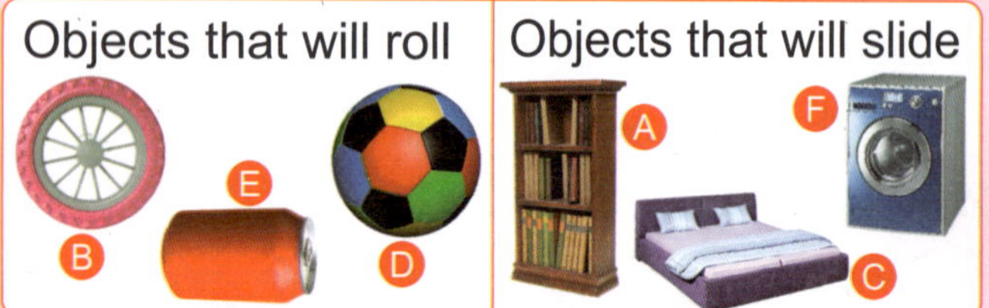